Demystifying Explosives: Concepts in High Energy Materials

S. Venugopalan
Former scientist,
High Energy Materials Research Laboratory,
Pune, India

ELSEVIER

AMSTERDAM BOSTON HEIDELBERG LONDON NEW YORK OXFORD
PARIS SAN DIEGO SAN FRANCISCO SINGAPORE SYDNEY TOKYO

A JOURNEY TOWARDS HIGHER EXPLOSIVE POWER

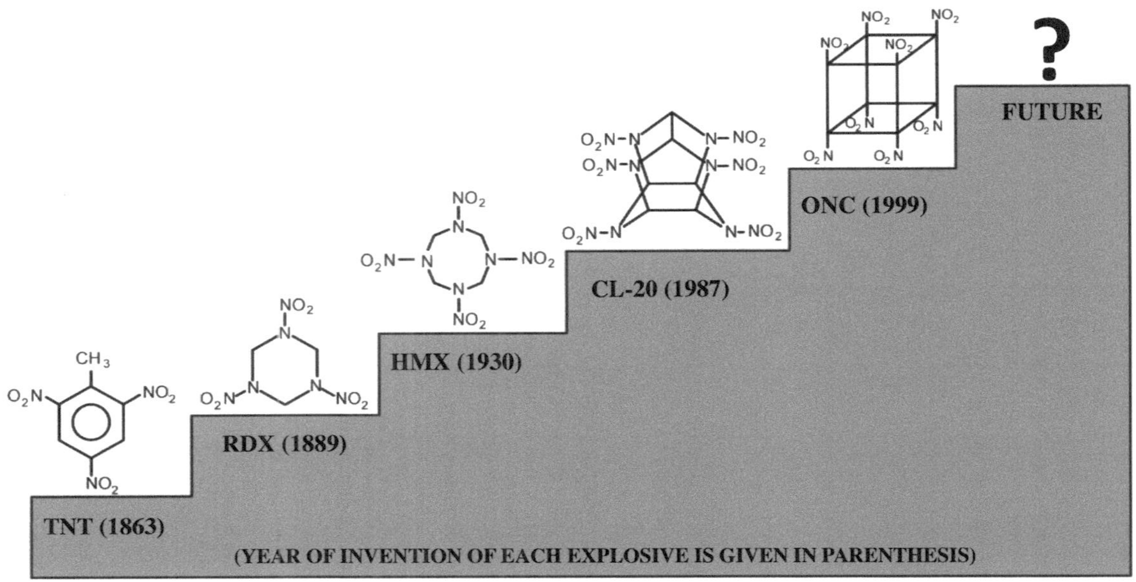

TNT – Trinitrotoluene

RDX - Cyclo trimethylene trinitramine
(**R**esearch & **D**evelopment **EX**plosive)

HMX - Cyclo tetramethylene tetranitramine
(**H**igh **M**elting **EX**plosive)

CL-20 – Hexanitrohexaazaisowurtzitane
(**C**hina **L**ake - 20)

ONC – Octanitrocubane

Elsevier
Radarweg 29, PO Box 211, 1000 AE Amsterdam, Netherlands
The Boulevard, Langford Lane, Kidlington, Oxford OX5 1GB, UK
225 Wyman Street, Waltham, MA 02451, USA

British Library Cataloguing in Publication Data
A catalogue record for this book is available from the British Library

Library of Congress Cataloging-in-Publication Data
A catalog record for this book is available from the Library of Congress

ISBN: 978-0-12-801576-6

Front cover photograph: Controlled explosion of unexploded ordnance. Courtesy: The U.S. Department of Defense *(DISCLAIMER: The use of military imagery does not imply or constitute endorsement of the author or his services by the U.S. Department of Defense)*

The author dedicates this book to the memory of countless number of innocent human beings who lost their lives in terrorist explosions all over the world with the prayer that this planet will be free from the clutches of terrorism in the near future.

High energy materials (HEMs; explosives, propellants, and pyrotechnics) are very dangerous if they are not handled with care and caution. In my career spanning over three decades, I have witnessed gruesome accidents, some of them fatal, at almost every stage, synthesis, scale-ups, production, testing, and even waste disposal of HEMs. The victims of those accidents include not only the beginners who were ignorant but also veterans who were either overconfident or complacent. *Odds are highly against you when HEMs are handled with disregard for standard operating procedures (SOPs) and the "Dos and Don'ts." Read Chapter 8 on safety for more details. Remember that like fire and electricity, HEMs can be your best friend or worst enemy depending on how you handle them.*

Contents

About the Author and Editor .... *xiii*
Foreword 1 .... *xv*
Foreword 2 .... *xvii*
Foreword 3 .... *xix*
Preface .... *xxi*
Acknowledgments .... *xxiii*
Abbreviations .... *xxv*

Chapter 1: In Pursuit of Energy and Energetic Materials *1*
- 1.1 Introduction 1
- 1.2 Gunpowder to Nitrocubanes 1
- 1.3 Classification of Explosives 6
  - 1.3.1 Maximum Power per Unit Volume 6
  - 1.3.2 High Velocity of Detonation 6
  - 1.3.3 Long-Term Storage Stability 7
  - 1.3.4 Insensitivity to Shock and Impact 7
  - 1.3.5 Ability to Withstand Large Accelerations 7
- 1.4 Explosives and Molecular Structure 8
- 1.5 Classification of Propellants 8
  - 1.5.1 Small-Arms Propellant 10
  - 1.5.2 Mortar Propellant 10
  - 1.5.3 Gun Propellant 10
  - 1.5.4 Rocket Propellant 11
- 1.6 Pyrotechnics 11
  - 1.6.1 Light 14
  - 1.6.2 Smoke 14
  - 1.6.3 Sound 14
  - 1.6.4 Heat 14
- Appendix A 15
- Appendix B 16
- Appendix C 16
- Suggested Reading 17
- Questions 17

Chapter 2: Energetics of Energetic Materials ... 19
2.1 Are Explosives and Propellants High-Energy Materials? ... 19
2.2 Explosive: The Wonderful Lamp ... 20
2.3 Thermochemistry and Explosive Energy ... 22
2.3.1 Heat of Reaction ... 23
2.3.2 Heat of Formation ... 23
2.3.3 Heat of Explosion (ΔH_e) and Heat of Combustion (ΔH_c) ... 27
2.3.4 Oxygen Balance ... 29
Worked Example 2.1 ... 32
2.3.5 Heat of Explosion: Dependence on Heat of Formation and Oxygen Balance ... 33
2.3.6 OB of Composite Explosives ... 34
2.3.7 Hazard Assessment from OB ... 35
2.3.8 Composition of Gaseous Products ... 35
2.3.9 Significance and Limitations of OB ... 36
2.3.10 Detonation Temperature/Flame Temperature ... 37
Worked Example 2.2 ... 39
2.3.11 Gas Volume ... 42
Worked Example 2.3 ... 42
2.3.12 The nRT Wonder ... 43
Worked Example 2.4 ... 44
2.3.13 Pressure of Explosion ... 45
2.3.14 Density ... 45
Summary of Important Terms ... 46
Suggested Reading ... 48
Questions ... 49

Chapter 3: Two Faces of Explosion: Deflagration and Detonation ... 51
3.1 Explosion ... 51
3.2 Deflagration and Detonation ... 52
3.3 Linear Burning and Mass Burning ... 54
3.4 Shock Wave and Detonation Wave ... 55
3.4.1 The Concept of a Shock Wave ... 56
3.4.2 Detonation Wave ... 58
3.5 Detonation Theory ... 60
3.6 Theoretical Estimation: VOD and P_d ... 62
3.6.1 Kamlet–Jacob Method ... 62
3.6.2 Becker-Kistiakowsky-Wilson Method ... 64
3.6.3 Rothestein and Petersen Method ... 65
3.6.4 Stine Method ... 67
3.7 Deflagration-to-Detonation Transition ... 67
3.7.1 When Can DDT Occur? ... 67
Suggested Reading ... 68
Questions ... 69

Chapter 4: HEMs: The Facet of Explosive Performance 71
4.1 Why Do Explosives Explode? 71
4.1.1 Spontaneity of Explosive Reactions 71
4.1.2 The Kinetic Aspect of Explosive Reactions 72
4.1.3 Molecular Structure and Explosive Properties 74
4.2 Two Aspects of Explosive Performance 77
4.2.1 Partition of Explosive Energy 77
4.2.2 Velocity of Detonation 78
4.2.3 Gas Expansion 79
4.3 A Travel by Explosive Train 81
4.3.1 Primary Explosives: "*the engine starters in an explosive train!*" 81
4.3.2 Secondary Explosives: "the final operators of the explosive train!" 83
4.3.3 The Types of Explosive Trains 86
4.4 Performance Parameters of Military Explosives 87
4.4.1 Fragmentation 87
4.4.2 Scabbing 88
4.4.3 Shaped Charge Penetration 89
4.4.4 Blast 91
4.5 Industrial Explosives 94
4.5.1 Introduction 94
4.5.2 Requirements of Industrial Explosives 95
4.5.3 Industrial High Explosives 96
4.5.4 Blasting Agents 97
4.6 Processing of the Compositions 100
4.6.1 Melt-Casting 100
4.6.2 Pressing 101
4.6.3 Plastic Bonded Explosives (PBX) 101
Suggested Reading 102
Questions 103

Chapter 5: The Propulsive Facet of HEMs: I (Gun Propellants) 105
5.1 Introduction 105
5.2 Gun: the Heat Engine 105
Worked Example 5.1 107
5.3 Unfolding Drama inside the Barrel 108
5.4 Energetics of Gun Propellant 110
Worked Example 5.2 112
5.5 Configuration of Propellant Grains 112
5.5.1 Regressive Burning 114
5.5.2 Neutral Burning 114
5.5.3 Progressive Burning 114
5.6 Salient Aspects of Internal Ballistics of Guns 116
Worked Example 5.3 117

5.7 The Chemistry of Gun Propellant Formulations 120
5.7.1 Role of Ingredients 123
Worked Example 5.4 125
Suggested Reading 131
Questions 131

Chapter 6: The Propulsive Facet of High Energy Materials—II (Rocket Propellants) *133*
6.1 Introduction to Rocketry 133
6.2 Basic Principles of Rocket Propulsion 133
6.2.1 Types of Rocket Engines 135
6.3 Specific Impulse 138
6.3.1 The Unit of I_{sp} 138
6.3.2 I_{sp} and Exhaust Velocity of Gases 139
Worked Example 6.1 139
6.4 Thermochemistry of Rocket Propulsion 140
6.5 Some Vital Parameters in the Internal Ballistics of Rockets 142
6.5.1 Linear Burning Rate 142
6.5.2 Characteristic Velocity 144
6.6 Design of a Rocket Propellant Grain 145
Worked Example 6.2 146
6.7 Chemistry of Solid Rocket Propellants 147
6.7.1 Choices and Limitations 147
6.8 Future of Rocket Propellants 153
Suggested Reading 154
Questions 154

Chapter 7: High Energy Materials in Pyrotechnics *157*
7.1 Introduction 157
7.2 Applications 157
7.3 Basic Principles of Pyrotechnics 159
7.3.1 The Chemical Components of Pyrotechnics 159
7.3.2 Factors Affecting the Performance of Pyrotechnics 161
7.3.3 Safety Aspects Involving Pyrotechnics 162
7.4 Conclusion 163
Suggested Reading 163
Questions 163

Chapter 8: HEMs: Concerns of Safety *165*
8.1 Introduction 165
8.2 Nature of Hazards 165
8.3 Hazard Classification of HEMs 166
8.4 The Damages 167
8.5 General Safety Directives 168
8.5.1 Assume the Hazard 168
8.5.2 Never Work Alone! 168

8.5.3 Start with the Smallest Possible Quantities 168
8.5.4 Safety Shields 168
8.5.5 Fire Hazards: Expect and be Ready 168
8.5.6 Ground (Earth) Your Facilities 168
8.5.7 Wear Protective Garments/Equipments (Including Antistatic Ones) 169
8.5.8 Practice Relative Humidity Control 169
8.5.9 Housekeeping 169
8.5.10 Know about the Material Hazards 169
8.5.11 Toxic Hazards 169
8.5.12 Prepare a Work Plan 170
8.5.13 Hazard Evaluation 170
8.5.14 Storage/Transport 170
8.5.15 Waste Disposal 172
8.6 Conclusion 172
Suggested Reading 172
Questions 172

Chapter 9: HEMs: Concerns of Security 173
9.1 HEMs: Concerns of Security 173
9.2 Detection of Explosives 174
9.2.1 Electron Capture Detector 175
9.2.2 Ion Mobility Spectrometer 176
9.2.3 Thermoredox Detector 176
9.2.4 Field Ion Spectrometer 177
9.2.5 Diamagnetism-Based Magnetic Field Detector 177
9.2.6 Nuclear Quadrupole Resonance Detector 177
9.2.7 Micro Electro Mechanical Systems 178
Suggested Reading 179
Questions 179

Chapter 10: HEMs: Characterization and Evaluation 181
10.1 Introduction 181
10.2 Chromatographic Techniques 182
10.2.1 Thin Layer Chromatography 182
10.2.2 Gas Chromatography 182
10.2.3 High Performance Liquid Chromatography 183
10.3 Spectroscopic Techniques 184
10.3.1 UV/VIS Spectroscopy 184
10.3.2 IR Spectroscopy 184
10.3.3 Nuclear Magnetic Resonance Spectroscopy 185
10.4 Thermal Evaluation of Energetic Materials 186
10.4.1 Differential Thermal Analysis 187
10.4.2 Differential Scanning Calorimetry 187
10.4.3 Thermogravimetric Analysis 189
10.4.4 Simultaneous Thermal Analysis 191

10.5 Sensitivity Tests of HEMs 191
10.5.1 Impact Sensitivity 192
10.5.2 Friction Sensitivity 192
10.5.3 Spark Sensitivity 192
Suggested Reading 193
Questions 193

Chapter 11: HEMs: Trends and Challenges 195
11.1 Introduction 195
11.2 Primary Explosives 195
11.2.1 Problems 195
11.2.2 Solutions 195
11.3 High Explosives 196
11.3.1 High-Density, High-VOD Explosives 197
11.3.2 Insensitive High Explosives 199
11.4 Propellants 202
11.4.1 Ecofriendly Oxidizers 202
11.4.2 Metallic Fuels 204
11.4.3 Energetic Binders 205
11.4.4 Thermoplastic Elastomers 206
11.4.5 Energetic Plasticizers 207
11.5 Polynitrogen Cages: Promising a Revolution in Future HEMs? 207
Suggested Reading 209
Questions 209

Chapter 12: HEMs: Constructive Applications 211
12.1 HEMs Have Shaped Our World 211
12.1.1 Mining and Quarrying 211
12.1.2 Construction 211
12.1.3 Oil Well Perforation 212
12.2 Controlled Demolition 212
12.2.1 Explosion or Implosion? 212
12.2.2 Step-by-Step 213
12.3 Air Bags 214
12.4 Explosive Welding 216
12.5 Avalanche Control 217
12.6 Life Saving Applications 218
Suggested Reading 218
Questions 219

Index 221

About the Author and Editor

S. Venugopalan did his postgraduation in chemistry from St Joseph's College, Tiruchirapalli (University of Madras). After teaching chemistry for 5 years, he served as quality assurance officer in a factory manufacturing a variety of explosives and propellants for small arms, guns, and rockets. Later he joined High Energy Materials Research Laboratory (HEMRL), Pune as a scientist and worked in the field of composite propellants and synthesis of energetic oxidizers and polymeric binders. He was also heading the Safety Engineering Division of the laboratory for about 6 years. His long experience and exposure to different types of HEMs in production, quality assurance, and research and development totals to about 32 years.

R. Sivabalan received his PhD in chemistry from Anna University, Chennai. He worked in the field of synthesis of advanced HEMs and insensitive munitions at HEMRL. He served as a postdoctoral researcher at Nanyang Technological University, Singapore. He has published about 40 research papers and has filed 3 patents and a copyright. Presently, he is working at Combat Vehicles Research & Development Establishment, Chennai.

Foreword 1

Condensed Systems Combustion Laboratory
V.V. Voevodsky Institute of
Chemical Kinetics and Combustion

Russian Academy of Sciences, Siberian Branch

3, Institutskaya Str., Novosibirsk 630090, Russia

There are a large number of books and reports that give copious information on explosives, propellants, and pyrotechnics which are grouped as high energy materials (HEMs) or energetic materials. But, the book by S. Venugopalan is probably the first book of its kind intended to make the basic concepts of different HEMs clear and highlight the intricate relationship between them. The book also touches upon various aspects of HEMs like concerns of safety and security, instrumental characterization and performance evaluation of HEMs, future trends, and also their constructive applications. With the help of this book, the author wants to "demystify" the notion on explosives (a feared word!) and popularize the field of HEMs right from college curriculum to the beginners as well as veterans working in this field!

The author has a vast practical experience in the fields of propellants, explosives, and synthesis of HEMs and related materials, apart from close interaction with their manufacture, quality assurance, and safety. Therefore, he makes it interesting to read this book by reporting problems of practical interest and possible solutions. The Russians would refer this as obtaining information "from the first hands." The author's style of presentation is amazingly simple and attractive. The book contains original statements/explanations regarding the definition and classification of HEMs and also their interrelationship in terms of energetics. The clearly illustrated pictures, worked examples, questions at the end of each chapter, and the suggested books for reading will be very useful to get a deeper understanding of the concepts.

The chapter on rocket propellants explains the fundamentals of internal ballistics of rockets and their bearing on the chemistry of formulation of the propellant as well as the challenges faced by the propellant chemist. There is a huge future for rocketry with many ambitious space programs in many countries and a youngster who dreams of choosing rocket propellants as his career will benefit from this chapter to get an introduction. While air-breathing engines like ramjets make use of the atmospheric oxygen for fuel combustion, the rockets that soar to outer space, where there is no oxygen, have to depend on compositions based on HEMs which provide an optimum combination of fuel and oxidizer.

The future of the rocketry can be fantastic when advanced HEMs like polynitrogen compounds can be used as propellant ingredients. Being highly endothermic compounds, these will release huge amounts of energy on decomposition resulting in very high values of specific impulse for the rockets, though many practical challenges may be encountered to use them, like safety, cost, and combustion stability.

I strongly believe that the information incorporated in the book will be quite useful to students, researchers, scientists, and technologists in understanding the basic and fundamental concepts of energetic materials. The contents and structure of the book are brilliant, and for the same reason, this will be an excellent material for teaching.

(Prof.V.E.Zarko)

Vladimir Zarko received his PhD and DSc from the Institute of Hydrodynamics, Novosibirsk, in 1985 and became Professor in Novosibirsk Technical University in 1989. He got several medals of Russian Federation of Cosmonautics for applied research and students' education. He has published five books and more than 150 papers in the field of energetic materials and holds 11 patents. In 1993, he was elected Honorary Member of HEMSI, India, and in 1997, Associate Fellow, AIAA, USA. He was invited researcher in Illinois University in 1993–1994 and in California University in Berkeley in 1997. In 2012, he taught the combustion course in Technion (Israel Institute of Technology), Haifa, Israel.

Vladimir E. Zarko
Professor and Head

Foreword 2

There are several books covering many different aspects of energetic materials, such as explosives, propellants, and pyrotechnics. Many of these examples go into great depth and detail. However, there are very few books available that cover, in a general way, the main concepts associated with high energy materials (HEMs). This book is one of the first that ties together many of the subjects important to understanding HEMs from a broad perspective.

This book covers topics at a conceptual level and help the reader obtain a good foundation. Examples of topics covered include: energetics of energetic materials, deflagration vs detonation, performance, propulsion, pyrotechnics, safety and security issues, characterization and evaluation, trends and challenges, and applications.

The book also provides many examples of problems that are solved in step-by-step detail to help the reader obtain a good understanding of subject matter being covered. Each chapter ends with a presentation of questions that cover the main concepts as well as references and suggested reading. The chapters are also written in a very clear manner and S. Venugopalan does an excellent job explaining the many diverse and difficult concepts associated with HEMs.

This book will be very beneficial to people who work in all different areas of energetic materials, and will be particularly useful for beginners in the field. The book will allow workers in energetic materials to understand how each HEM concept relates to one another. The book will be an excellent addition to not only the libraries meant for HEMs like propellants, explosives, and pyrotechnics, but also to universities and college libraries, so that a scientific awareness about HEMs can be spread among students with a chemistry background.

Dr. David Chavez, Scientist, Los Alamos National Laboratory,

New Mexico, United States of America

David Chavez received his BS with honor in chemistry from the California Institute of Technology and PhD from Harvard University. He was a National Science Foundation and Beinecke Memorial predoctoral fellow, a Frederick Reines Distinguished Fellow at Los Alamos National Laboratory, and is an invited Professor at the Ecole Normale Superieure, in Cachan, France.

In 2011, he was awarded the prestigious E. O. Lawrence Award in the Atomic and Molecular Sciences category. He has published over 50 papers in the areas of organic chemistry and energetic materials synthesis (with over 1800 citations) and holds 10 patents in energetic materials and pyrotechnics.

Foreword 3

I feel privileged to write a foreword to this book, authored by S. Venugopalan who was my senior colleague in HEMRL for a number of years. I can recall that with his rich experience in the field of high energy materials (HEMs) and a strong background in fundamental and applied chemistry, many of the scientific officers and staff would approach him to clarify their doubts in this field, be it propellants or explosives or the synthetic organic chemistry related to HEMs. He was a popular teacher and invited speaker on various subjects, particularly on HEMs and there was a growing request from the scientific fraternity of the laboratory that he should write a popular book mainly focusing on the basic concepts governing HEMs, their development, and applications. This book is the result of such a request and hard work by the author.

The author gives a lucid elucidation of some basic terms such as explosives, HEMs, deflagration, detonation, etc., with examples and also the classification of HEMs. His explanation of the energetic aspects of HEMs based on thermochemistry, especially the significance of the heats of formation of HEMs is quite original and outstandingly clear. The network chart depicting the interrelationship between different parameters of HEMs, at the end of Chapter 2 excellently sums up the basic concepts of HEMs. As the author makes it clear in the preface, this book is meant mainly to create an interest in the field of HEMs among the beginners. A college student with a degree in chemistry can easily understand the intricacies related to explosives, propellants, and pyrotechnics and can be motivated to choose HEMs as his/her career. The book touches upon all the aspects concerning HEMs including safety and security concerns, instrumental analysis for their characterization and performance evaluation, future trends, and interestingly, the constructive applications of HEMs. The worked examples of numerical problems in quite a few chapters and the questions at the end of each chapter should be useful to the readers.

In light of my above comments, I strongly feel that this book should find a place not only with every scientist and technologist working in institutions handling HEMs but also in the libraries of colleges teaching chemistry to enhance the awareness about the importance and scope of HEMs. Apart from the beginners, even an experienced researcher in the field of HEMs will find this book an asset as he will understand the broader perspective of the entire gamut of HEMs that will help him in his work. I am confident that the book will be a unique **popular science publication** with the hope that **HEMs chemistry** may become a part of the chemistry curriculum in many universities and colleges, like other branches of chemistry, in the near future.

(Dr.Mahadev B.Talawar), Scientist,
High Energy Materials Research Laboratory (HEMRL), Pune, India.

Dr Talawar was awarded PhD from Karnataka University, India, in 1994. He has been working in the indigenous development of advanced HEMs of defense interest for two decades. He has authored/coauthored nearly 150 research papers in the area of materials science in the peer-reviewed national and international journals of repute. He has presented several research papers in national and international seminars in the area of HEMs. He was a visiting scientist at Mendeleev University of Chemical Technology, Moscow, Russia, during 1998. Dr Talawar is also serving as an Editorial Board member for reputed journals such as Journal of Hazardous Materials, USA and Combustion, Explosion and Shockwaves from Russia. He has also been reviewing research papers in the area of materials science for many international journals. Dr Talawar worked as a Senior Chemical Weapons Inspector for the Organization for the Prohibition of Chemical Weapons (OPCW) at the Netherlands during 2005–2012. During this period, he acquired unique experience in the specialized field of destruction of chemical weapons. As a part of OPCW, he has visited about 50 countries and immensely contributed to various inspection activities.

Preface

The history of explosives dates back to more than 2000 years and it is a matter of common knowledge that Chinese were the first to make the first ever "explosive," namely, gunpowder or black powder sometime before 200 BC. There was a huge lull in the field for nearly 1400 years since then, till Roger Bacon, an English monk carried out detailed experiments on black powder around AD 1249. But, the real momentum in the development of explosives and propellants picked up only in the midpart of the nineteenth century with a number of contributors, mostly from Europe, Alfred Nobel being the most notable among them. A gist of the important milestones in the development of explosives and propellants is given in Chapter 1.

Twentieth century has witnessed some remarkable milestones in the synthesis of explosives of high power, higher thermal stability, and low vulnerability. Simultaneously, great progress was made in the development of propellants for rockets, guns, mortars, and small arms. Similar milestones were reached in the field of pyrotechnics which are essential parts of any system that uses explosives and propellants. Many major breakthroughs in the field of explosives, propellants, and pyrotechnics (collectively and loosely named as "high energy materials" (HEMs)) were possible in the twentieth century because of great strides that were made in the fields of chemistry—particularly synthetic organic chemistry, advanced instrumentation, detonics, and engineering. Despite the impressive progress witnessed in the field of HEMs, during the last century, it must be admitted that the rate of progress is much slower as compared to other fields like polymer chemistry, electronics, and computers owing to a number of constraints and restrictions an HEM scientist has to encounter in developing a new HEM, like safety, stability (thermal, mechanical, storage, etc.), cost, and other considerations.

Excellent books, manuals, and journals are available in the field of HEMs (important journals mentioned at Chapter 1) and with the advent of the Web, large amount of information on HEMs is only a click away. But I felt that there is a need for a book where the main thrust will be on the various **CONCEPTS** of HEMs rather than details of their preparation, properties, and applications. With about more than 30 years of experience in HEMs, having been associated with production, quality assurance, and R&D related to explosives and propellants of various types, I realized that there exists a need for a book with the main purpose of making the basic concepts of HEMs clear for the HEM community as a whole. This book is the result of that realization wherein I have tried to illustrate the concepts in as simple manner as possible so that the reading becomes easy, interesting, and assimilable. I hope that this book will be particularly useful to the beginners in the field of HEMs, whether they are in production or inspection or R&D.

It is possible that this being the first edition, there can be errors or commissions or omissions at some places. In such cases, I will be grateful if they are brought to my notice along with any constructive suggestions so that the necessary corrections/editing can be done in the next edition.

Acknowledgments

I wish to acknowledge and thank the following persons who helped me in bringing this book to completion: **Dr R. Sivabalan**, an experienced researcher in the field of synthesis of explosives for having agreed to edit this book, **Dr H. S. Yadav**, retired scientist from HEMRL, Pune, for his inputs regarding the discussion on detonics and shockwaves, **Dr Harries Muthurajan** and **Ms Marine** for the technical support in typing and formatting the original manuscript, **Mr Vijay Venugopalan,** my son**,** for all the help and support he has given me to complete the book, and also the **scientists and staff of HEMRL, Pune**, who spurred me to write this book.

I am grateful to **Prof. Vladimir Zarko**, Head of the Institute of Chemical Kinetics, Novosibirsk, Russian Academy of Sciences, Russia, **Dr David Chavez**, Los Alamos National Laboratory, USA, and **Dr M. B. Talawar**, HEMRL, Pune, India, for their review of the book followed by constructive suggestions.

S. Venugopalan

Abbreviations

ADN: Ammonium dinitramide
AMATOL: Ammonium Nitrate (40%) and Trinitro toluene (60%) mixture
AN: Ammonium nitrate
ANFO: Ammonium Nitrate Fuel Oil
AP: Ammonium Perchlorate
BAMO: Bis-Azido Methyl Oxetane
BDNPA: Bis-(2,2-dinitropropyl) Acetal
BDNPF: Bis-(2,2-dinitropropyl) Formal
BNCP: Bis-(5-nitro-2H-tetrazolato-N) Tetramine Cobalt (III) Perchlorate
BTATz: Bis-tetrazolyl Amino Tetrazine
Bu-NENA: Butyl-Nitrato Ethyl Nitramine
BTTN: Butane Triol Trinitrate
CD (nozzle): Convergent-Divergent (Nozzle)
CE: Composition Exploding (also called Tetryl)
CL-20: China Lake-20 (also called HNIW)
CTPB: Carboxy Terminated Poly-Butadiene
CYCLOTOL: RDX (77%) and TNT (23%) mixture
DBP: Dibutyl Phthalate
DDT: Deflagration-to-Detonation Transition
DMNB: 2,3-Dimethyl-2,3-Dinitro Butane
DNAN: 2,4-Dinitro Anisole
DNB: Dinitro Benzene
DNT: Dinitro Toluene
DOP: Dioctyl Phthalate
DPA: Diphenyl Amine
2N-DPA: 2-Nitro Diphenyl Amine

DSC:	Differential Scanning Calorimetry
DTA:	Differential Thermal Analysis
ECD:	Electron Capture Detector
EGDN:	Ethylene Glycol Dinitrate
ESD:	Electrostatic Discharge
ESH:	Explosive Storage House
FIS:	Field Ion Spectrometer
FOX-7:	1,1-diamino-2,2-dinitroethylene
GAP:	Glycidyl Azide Polymer
GC:	Gas Chromatography
HD:	Hazard Division
HE:	High Explosive
HEAT:	High Explosive Anti-Tank
HEM:	High Energy Material
HESH:	High Explosive Squash Head
HMTD:	Hexa Methylene Triperoxide Diamine
HMX:	High Melting Explosive (also called Octogen): Cyclotetramethylene tetranitramine
HNF:	Hydrazinium Nitroformate
HNS:	Hexanitro Stilbene
HNIW:	Hexanitro hexaaza Isowurtzitane (also called CL-20)
HPLC:	High Performance Liquid Chromatography
HTPB:	Hydroxy Terminated Poly-Butadiene
IED:	Improvised Explosive Device
IM:	Insensitive Munitions
IMS:	Ion Mobility Spectrometer
IR:	Infra Red
Isp:	Specific Impulse
LA:	Lead Azide
LBR:	Linear Burning Rate
LLM-105:	2,6-diamino-3,5-dinitropyrazine-1-oxide (a thermally stable insensitive explosive) developed by Lawrence Livermore National Laboratory, USA.
LOVA:	Low Vulnerability Ammunition

LOVEX:	Low Vulnerability Explosive
LOX:	Liquid Oxygen
MEMS:	Micro Electro Mechanical System
MF:	Mercury Fulminate
MNT:	Mononitro Toluene
MSDS:	Materials Safety Data Sheet
MTNI:	N-Methyl-2,4,5-Trinitro Imidazole
NC:	Nitrocellulose
NEQ:	Net Explosive Quantity
NG:	Nitroglycerine
NHN:	Nickel Hydrazine Nitrate
NIMMO:	3-Nitratomethyl-3-methyl oxetane
NMR:	Nuclear Magnetic Resonance
NTO:	3-Nitro-1,2,4-triazole-5-one
OB:	Oxygen Balance
OCTOL:	HMX (76.3%) and TNT (23.7%) mixture
ONC:	Octanitro Cubane
OQD:	Outside Quantity Distance
PBX:	Plastic Bonded Explosive
PETN:	Pentaerythritol Tetranitrate
PGN:	Poly-Glycidyl Nitrate
PIQD:	Process Inside Quantity Distance
PVC:	Polyvinyl chloride
RDX:	Research & Development Explosive (also called Hexogen): Cyclotrimethylene trinitramine
RFNA:	Red Fuming Nitric Acid
SEMTEX:	Generic name for RDX-based plastic bonded explosive
SIQD:	Storage Inside Quantity Distance
SOP:	Standard Operating Procedure
STA:	Simultaneous Thermal Analysis
TATB:	Triamino Trinitro Benzene
TATP:	Triacetone Triperoxide
TACOT:	Tetranitro dibenzo tetraazapentalene (a heat resistant explosive)

TAGAT: Triamino Guanidinium Azotetrazolate
TDI: Toluene Di-isocyanate
TEGDN: Triethylene Glycol Dinitrate
Tetryl: 2,4,6-trinitrophenyl methyl nitramine (also called CE)
TGA: Thermogravimetric Analysis
THF: Tetrahydrofuran
TLC: Thin Layer Chromatography
TMD: Theoretical Maximum Density
TMETN: Trimethylolethane trinitrate
TNAZ: 1,3,3-Trinitro Azetidine
TNB: Trinitrobenzene
TNT: Trinitrotoluene
TORPEX: Explosive Mixture of RDX (40.5%), TNT (40.5%), Aluminum (18%), and Wax (1%)
TPE: Thermoplastic Elastomer
TRD: Thermoredox Detector
TRITONAL: Explosive Mixture of TNT (80%) and Aluminum (20%)
UV: Ultraviolet
VOD: Velocity of Detonation

CHAPTER 1

In Pursuit of Energy and Energetic Materials

1.1 Introduction

In Hindu mythology, "energy" has been given a place of pride. Similar to Greeks who deified the qualities of love and valor, Hindus deify energy (*Shakti*) as Goddess *Kali*. The existence of life on the Earth is unthinkable in the absence of sources of energy and energy-giving materials. Since the evolution of human civilization, man has been in tireless pursuit of sources that provide him more energy for livelihood, comfort, and advancement. Evidently, the first "energetic" material that the prehistoric man used was firewood that burned (or underwent combustion in a more scientific parlance) to provide him the source of heat with which he could cook meat and vegetables for more palatable consumption. It is interesting to note that since the commencement of civilization, until a few centuries back, firewood was the main fuel for providing energy to man.

The discovery of coal helped him to make giant leaps in the process of industrial advancement. With the advent of oil hardly two centuries back, the very pattern of life all over the world has radically changed. Today, oil is the lifeline of modern living. Despite the possibility of using nuclear energy and other nonconventional sources of energy, such as solar energy, tidal energy etc., oil still rules the roost and one is justifiably worried about what would happen, say, after a century or so when the indiscriminate tapping of this fossil fuel from the mother earth will leave our posterity high and dry.

1.2 Gunpowder to Nitrocubanes

The so-called energetic or energy-giving materials mentioned in the preceding paragraph viz. firewood, coal, and oil are actually fuels. Unless oxygen from air is available to them, they do not burn and give the energy in the form of heat. However, man, who, with his inborn aggressive instinct has caused several wars, was not to be satisfied with fuels such as the above, which he thought could be reserved only for cooking, illumination, and other similar activities. To advance from the arrow-bow-spear-sword warfare, he wanted something that would propel a harmful projectile, preferably through a barrel, at his

Demystifying Explosives: Concepts in High Energy Materials. http://dx.doi.org/10.1016/B978-0-12-801576-6.00001-X

enemy. The first material that met such a need was gunpowder, which, as we all know today, is a physically intimate mixture of finely ground potassium nitrate (KNO_3; 75%), charcoal (carbon, 15%), and sulfur (10%). Here was something that did not depend on atmospheric oxygen because most of the needed oxygen for combustion came from oxidizer KNO_3. It was in the fourteenth century that the monk Berthold Schwarz invented a gun and used black powder for propelling stones from it. This discovery of the usefulness of black powder for accomplishing mechanical work may be considered as the real beginning of the history of explosives. It was only required to simply load the gunpowder in the cannon and ignite it so that the high-pressure gases produced would propel the cannon balls on to the enemy. The various decomposition reactions of gunpowder are as follows:

$$2KNO_3 + 3C + S \rightarrow K_2S + N_2 + 3CO_2$$

$$4KNO_3 + 7C + S \rightarrow 3CO_2 + 3CO + 2N_2 + K_2CO_3 + K_2S$$

More than 1000 years back, Chinese appeared to have invented gunpowder mainly for the purpose of fireworks. During the year 1250, Roger Bacon described the constituents of gunpowder, but its first use in guns was made by the British in the year 1346 at the Battle of Crecy. Many wars of the eighteenth and nineteenth centuries are reported to have been fought with gunpowder playing a key role. Rockets using gunpowder were used against Napoleon's army between 1803 and 1815. Tippu Sultan, a king from South India, is reported to have inflicted heavy casualties on the British Army in several battles using gunpowder. We should note that in all of these cases, the gunpowder was used only as a propulsive material.

The explosive property of gunpowder was reported by Roger Bacon in the thirteenth century and was rediscovered by Shwarz in Germany in the fourteenth century. During the seventeenth century, the explosive property of gunpowder was used for quarrying purposes in Europe. It must be remembered that gunpowder is basically a "deflagrating" (i.e., fast, layer-by-layer burning) material and at certain conditions (e.g., confinement), the deflagration gets converted into violent "detonation" (i.e., explosion accompanied by destructive shock wave).

The use of gunpowder was a messy affair. It was dirty to handle; it fouled the gun barrels; it was unpredictable in performance; and, above all, it produced so much smoke and flash that the enemy had no problem in locating the position of the gun. Therefore, there was a search to make a "smokeless propellant." One route was to prepare compounds in which a single molecule had the "oxidizer" element, "fuel" element, and the energy-giving moiety. There would not be any need for mixing oxidizers and fuels to achieve propulsion. In the middle of the nineteenth century, chemistry had considerably advanced in Europe, and the chemists concentrated their efforts to produce compounds

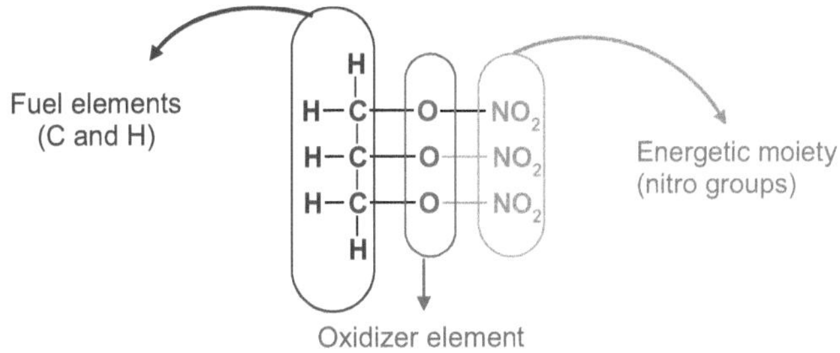

Figure 1.1
Molecule of Nitroglycerine (NG).

that would satisfy this three-in-one requirement. They nitrated several organic compounds to get their nitrato or nitro products. Nitrocellulose (NC), nitroglycerine (NG), and trinitrotoluene (TNT) were some of the products that could serve their purpose. For example, in a molecule of NG (obtained by the nitration of glycerine), we have the fuel elements, oxidizer element, and the energy-giving (or energetic) moiety viz. nitro groups as shown in Figure 1.1.

[The nomenclature "nitroglycerine" is a misnomer. Basically, NG contains three nitrate ($-O-NO_2$) groups and should be actually called "glyceryl trinitrate," produced by the esterification of glycerol by nitric acid (HNO_3). Likewise, NC is, in fact, cellulose nitrate].

NG is prepared by slowly adding glycerol to a mixture of HNO_3 and sulfuric acid (H_2SO_4) maintained at 20 °C.

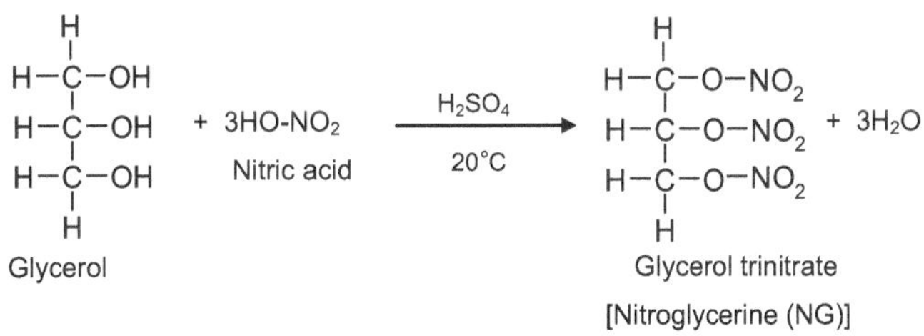

The NG molecule does not depend on external oxygen. On the other hand, it has a slight excess of oxygen after oxidation of carbon and hydrogen to carbon dioxide (CO_2) and water (H_2O), respectively.

$$4C_3H_5(NO_3)_3 \longrightarrow 12CO_2 + 10H_2O + 6N_2 + O_2 + \text{heat}(1467\ \text{kcals})$$

Such a substance that has excess of oxygen after combustion is said to have a positive oxygen balance.

One of the earliest explosives synthesized was NC (see Figure 1.2). During the 1830s, NC was synthesized by the nitration (using a mixture of conc. HNO_3 and conc.

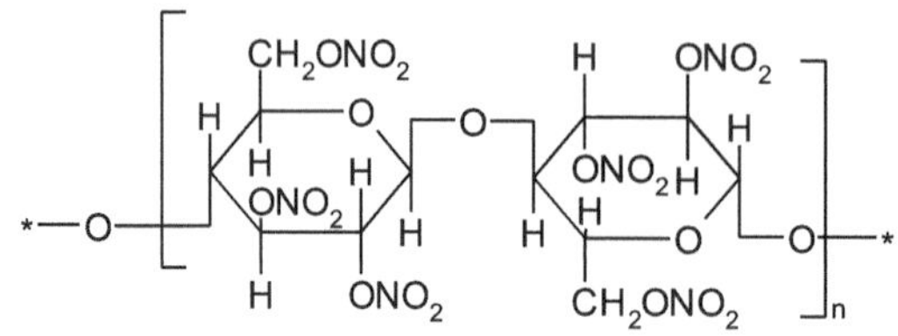

Figure 1.2
Structure of Nitrocellulose (NC).

H_2SO_4) of cellulose, a well-known natural polymer and a main constituent in plants. *NC is a fibrous high explosive and is dangerously sensitive in dry conditions.* NC needs to be stabilized after its preparation by proper chemical treatment. Otherwise, traces of acids and other byproducts accelerate its decomposition. In the middle of the nineteenth century, several disastrous explosions occurred in factories and storage houses because of the unstabilized NC. In 1866, Abel published his research work on the stabilization of NC. In 1884, the French scientist Vielle "gelatinized" NC by partly modifying its fibrous nature using a mixture of ether and alcohol so that it could be used as a reasonably insensitive "propellant powder" for manufacture and handling.

Synthesis:

$$\underset{\text{Cellulose}}{C_6H_7O_2(OH)_3} + \underset{\text{Nitric acid}}{3HNO_3} \xrightarrow{H_2SO_4} \underset{\text{NC}}{C_6H_7O_2(NO_3)_3} + 3H_2O$$

On explosion, NC gives carbon monoxide (CO), CO_2, H_2O, nitrogen, and heat as follows:

$$2C_6H_7O_2(NO_3)_3 \longrightarrow 9CO + 3CO_2 + 7H_2O + 3N_2 + \text{heat}$$

The modern era in the history of explosives began in 1838 with the preparation of NC by Pelouze by nitrating paper-based cellulose, but it was not until 1846 that its explosive properties were known. Similar to black powder, it was at first used for propellant and blasting purposes. In 1847, Ascanio Sobrero prepared NG, a powerful liquid explosive. NG was found to be extremely sensitive to shock, and Sobrero probably thought that the best thing was to put it under a lid and forget about it. However, after approximately 15 years, Alfred Nobel (1833–1896), the Swedish scientist, a prolific inventor and a philanthropist, opened that lid and started his struggle with NG. Despite its great hazard, he had the vision to realize its great potential and unshakeable confidence that one day he would tame it. (Alfred's tenacity was unbelievable. In an accident related to his work on detonators, his younger brother was killed and his father died heart-broken. However, Alfred never relented and carried on with his research and development in detonators). NG (nitroglycerine) is a dangerously sensitive material when it undergoes even mild impact. Kieselghur was found to be the first inert material, which when mixed with NG could desensitize NG, by taming its sensitivity. Dynamite was the first substance using NG that could be safely and conveniently handled. After a lapse of approximately 135 years, dynamite is still used for certain civil applications. A more startling discovery by Nobel followed when he mixed sensitive NG with sensitive NC and got an insensitive, gelatinized dough. This gelatinized material was found to be a powerful blaster and accordingly called "blasting gelatin." This discovery paved the way for the development of many blasting explosives (dry NC can absorb up to 11.5 times its weight of NG; accordingly, blasting gelatine has a composition of 92% NG and 8% NC). In 1888, Nobel developed the first "smokeless powder" (called ballistite) for military application in place of gun powder. It was a mixture of NC and NG (called "double-base") and substances such as camphor, which acted as plasticizers. Prolific inventions by Nobel resulted in many patents to the credit of this genius who revolutionized the explosives industry. He accumulated a huge fortune and it is common knowledge that the prestigious Nobel Prizes are given as a result of this fortune.

From the middle of nineteenth century, many explosives and energetic ingredients have been synthesized. Appendix A gives the important milestones in the history of development of explosives and propellants. During the last 150 years, with great strides made in the field of chemistry, physics, instrumentation, and computers, we have come a long way in the field of explosives and propellants. Scientists are constantly on the hunt for better and better candidate molecules with regard to energy content, rate of release of energy, density, and other parameters so that they can be used for futuristic explosive compositions or as propellant ingredients.

The pursuit that started with gunpowder some centuries back is still very much on. The targets are presently molecules that have a highly strained structure, have maximum density, and contain energetic groups. One such molecule that has been recently synthesized is octanitrocubane (Figure 1.3).

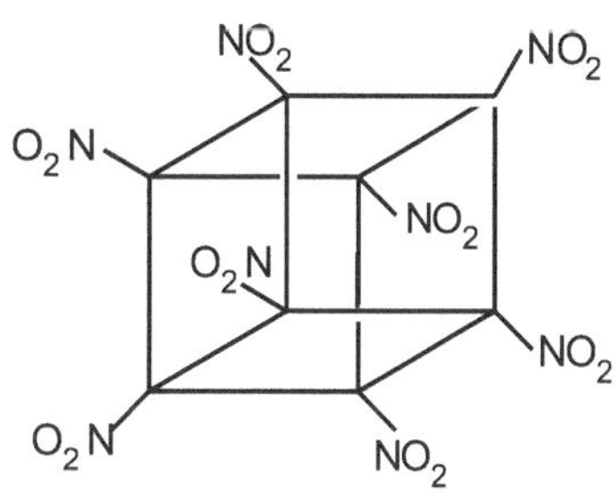

Figure 1.3
Octanitrocubane.

The advancing techniques of modern warfare lead to more and more specialized requirements for explosives and propellants. We will see in greater detail about such requirements in subsequent chapters.

1.3 Classification of Explosives

The term "explosives" has been defined in various ways and probably the most accepted definition is as follows: "An explosive is a chemical or mixture of chemicals which, when suitably initiated, can react so rapidly and with such liberation of energy that there can be damage to the surroundings." The explosives can be classified in two different ways:

1. High explosives or detonating explosives (subdivided into primary explosives and secondary explosives)
2. Low explosives or deflagrating explosives (propellants)

Propellants that propel a projectile from a pistol, mortar, or gun fall under the category of low explosives. They deflagrate layer by layer at a predetermined rate evolving a large amount of high-pressure hot gases that do the trick of propelling the projectile. Rocket propellants that cause the self-propulsion of a rocket are also referred to as low explosives.

Primary explosives are quite sensitive to initiation by mechanical impact, flame, or spark. Among high explosives, secondary explosives such as TNT and RDX (research and development explosive) detonate with violence, causing a high-velocity shock wave and blast effect. They are fairly insensitive and need to be initiated by primary explosives such as certain metallic azides. Although they are not as powerful as secondary explosives, primary explosives have enough power to detonate a secondary explosive on initiation.

The second type under which explosives can be classified is based on their end use viz. military explosives and civil explosives (sometimes referred to as commercial explosives). The requirements, properties, and cost are quite varied between these two categories of explosives. The important requirements of *military explosives* are presented in the following subsections.

1.3.1 Maximum Power per Unit Volume

This implies that a given volume of a high explosive (e.g., in a shell or warhead), on detonation, should produce high-pressure and high-temperature gases in such a way that they do maximum work (of expansion) per unit time.

1.3.2 High Velocity of Detonation

Velocity of detonation (VOD) is the rate at which the shock wave front travels in the medium of an explosive. This is a vital parameter for most of the military explosives because it is directly proportional to the shattering effect of the detonation (e.g., in a grenade) and the jet velocity in shaped charges. It is measured in units of kilometers per second or meters per second (e.g., VOD of RDX is 8.850 km/s).

1.3.3 Long-Term Storage Stability

Wars are not an everyday affair. In times of peace, all of the explosives-filled ammunitions are under storage over long periods of time—sometimes for a few decades. For this purpose, one cannot choose an explosive that deteriorates in stability within a few years. Therefore, military explosives should have long-term stability over a wide range of temperatures (e.g., from −40 °C to +60 °C).

1.3.4 Insensitivity to Shock and Impact

Explosives should go off when they are supposed to go off (reliability), but they should not go off when they are not supposed to go off (safety). Hence, this is an important safety-related requirement for any military explosive at various stages of handling and transport.

1.3.5 Ability to Withstand Large Accelerations

The high explosives filled in ammunition have to encounter huge accelerations (e.g., in the bore of a gun accelerations up to 40,000 g may be experienced) or negative accelerations (e.g., in the penetration by an armor-piercing shell through an armor plate). It should be ensured that the HE (high explosive) filling should not be initiated by such acceleration or deceleration.

In sharp contrast to military explosives, civil explosives do not generally require high VOD or too high of a blast effect. In fact, a high-VOD explosive may prove to be disastrous in certain civil applications such as coal mining in which the high-velocity shock waves might result in adiabatic compression and the initiation of the marsh gas (methane) present in the coal mines. The "power" of the explosives or explosive compositions can be tailored by adjusting the composition to suit the needs.

Cost is a dominant factor in civil explosives. One cannot afford to produce a costly civil explosive simply because it will not sell. A coal-mining magnate may simply ask you "What is the cost of your explosive needed to mine 1 ton of coal in the required range of lump size?"

Unlike military explosives, the shelf life of civil explosives is not very critical. After all, they are fast-moving items, and one might talk of shelf life of 6 months or 1 year versus 20 or 25 years for a military explosive.

1.4 Explosives and Molecular Structure

Why are only some compounds explosives whereas many others are not? Only in some compounds, the molecules are

1. having a huge potential energy packed in them (which is related to their heats of formation, a topic that will be discussed in the Chapter 2);
2. in a *meta*-stable state, implying that they need only a small amount of trigger or activation energy to initiate them for the release of the dormant potential energy in a very short time.

Because this is basically a molecular phenomenon, a relationship does exist between the molecular structure and the explosive property.

It was proposed that an explosive molecule has in its structure certain groups that are responsible for their explosive property (e.g., $-ONO_2$, $-NO_2$, N-NO_2, $-ClO_4$, $-N_3$, etc.). These were called "explosophores" (similar to "chromophore" groups such as an azo group that imparts color to a dye). Paul W. Cooper in his book *Explosives Engineering* talks about four different substituent groups found in explosives viz.

1. Oxidizer contributor (e.g., $-ONO_2$, $-NO_2$, $-NF_2$)
2. Fuel contributor (e.g., alkyl, $-NH_2$, $-NH$)
3. Combined fuel-oxidizer contributor ($-ONC$: fulminate, $-NH$-NO_2: nitramines)
4. Bond-energy contributor (e.g., $-N_3$: azides), which contributes energy to the detonation process when their high-energy bonds are broken.

If we take the above combinations, there must theoretically be thousands of explosives (particularly organic) available to us. However, the actual number of explosives being used or pursued is severely restricted by several factors, such as thermal stability, sensitivity, chemical compatibility, toxicity, and explosive output of the finally synthesized explosive apart from the cost and feasibility of the synthesis.

In terms of molecular structure, the explosives can be classified as shown in Figure 1.4.

1.5 Classification of Propellants

Propellants are defined as low explosives or deflagrating explosives. Such a definition is rather loose because the roles of a deflagrating explosive and detonating explosive can be reversed depending on the conditions. A propellant detonates under very high confinement and a high explosive such as RDX can be made to silently burn as a propellant ingredient. However, in this book, the term "propellant" will refer to those materials that burn (without the help of external oxygen) layer by layer at a predetermined or predictable rate evolving high pressure and hot gases.

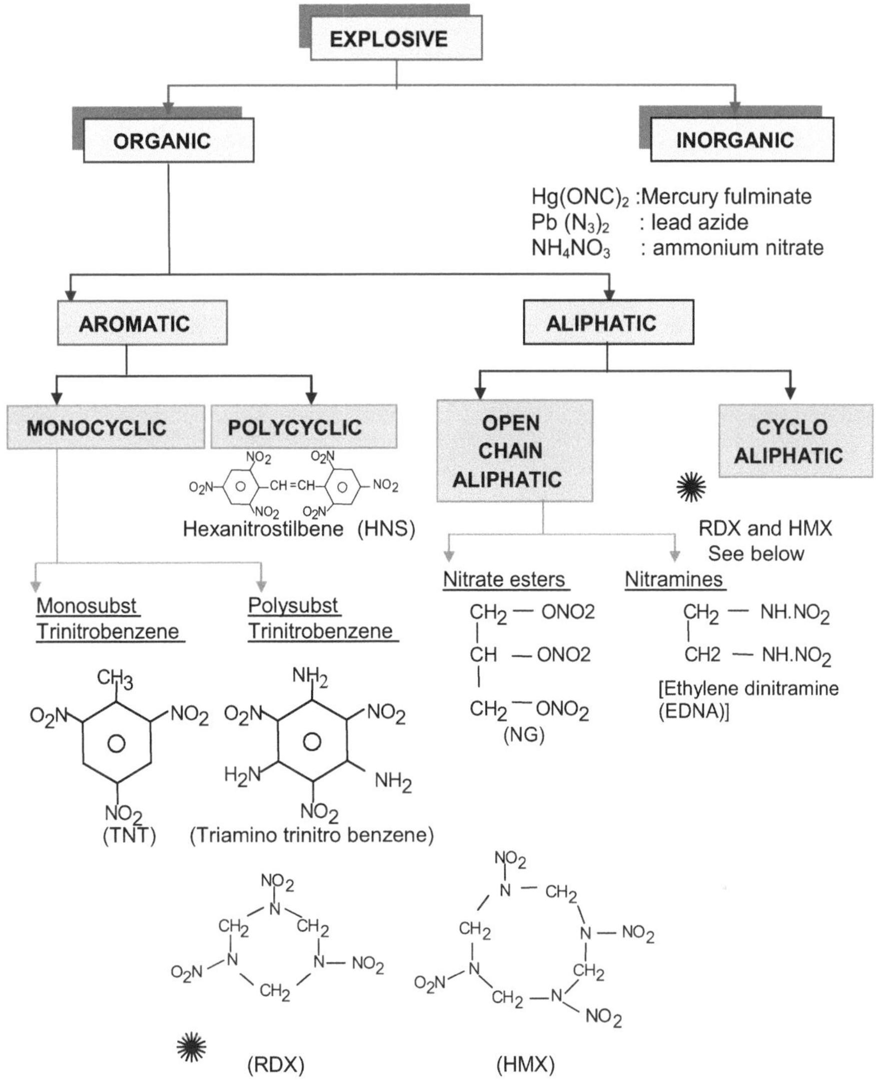

Figure 1.4

Classification of Explosives on the Basis of Their Molecular Structure.

Over the years, the field of propellants has grown vastly in terms of variety, application, and technology. The propellants can be classified based on either where they are going to be used (rockets or guns or small arms) or the chemical composition (i.e., single-base propellants containing mainly NC; double-base propellants containing NC and NG; and triple-base propellants containing NC, NG, and nitroguanidine; composite propellants containing a solid inorganic oxidizer such as ammonium perchlorate (NH_4ClO_4; AP) dispersed in a polymeric fuel-binder matrix; etc.). The chemistry and technology of these

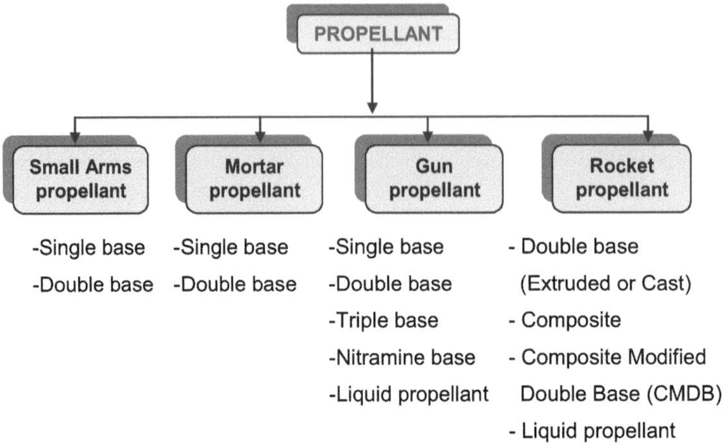

Figure 1.5
Classification of Propellants Based on Their End Use.

propellants will be discussed in subsequent chapters. A broad classification of propellants based on their end use is given in Figure 1.5.

1.5.1 Small-Arms Propellant

They are normally fine grains of propellants, mostly based on either single-base propellants (NC) or double-base propellants (NC + NG) including the so-called "ball powder." They are loaded in the cartridge cases meant for propelling bullets from rifles and pistols.

1.5.2 Mortar Propellant

Mortars are weapons used in warfare for propelling shells that are muzzle-loaded. Normally, there are two types of charges of propellant: primary and secondary. The most common composition is based on NC and NG, and the propellant is in the form of fine flakes of specified dimensions.

1.5.3 Gun Propellant

Gun propellants are also called "smokeless powders," a term that originated in the nineteenth century to distinguish the newly developed NC propellants from the traditional gunpowder. They are indeed largely smokeless on firing. Solid gun propellants mostly contain NC. In addition, conventional gun propellants consist of mixtures of one or more explosives with various additives, formulated and carefully processed to burn smoothly without detonating, under the conditions in which they are normally used. The essential required properties of gun propellants are as follows:

1. Minimal smoke or flash
2. Less toxic fumes
3. Long shelf life under all environmental conditions
4. Easy and rapid ignition
5. Low sensitivities to all other possible cause of initiation
6. Low flame temperature

1.5.4 Rocket Propellant

Rocket propellants provide a simple and effective way of creating propulsion for flight. The first true military use was by British troops in the eighteenth century against Indians. By 1805, Sir William Congreve had devised a system for the British, and these propellants provided an important military advantage in the following decade. By the start of World War I, such rockets, all powered by gunpowder, had become obsolete. Since then, the vital importance of rocket-powered weapons to attack on land, sea, and in the air has tremendously increased.

Basic information about the nature of composition and the application of various types of propellants is given in Table 1.1. The significance of the ingredients with respect to their chemistry and energetics vis-à-vis the final application will be discussed in the chapters on gun and rocket propellants.

1.6 Pyrotechnics

("Pyro" in Greek means "fire.")

The display of fireworks on festive occasions has been recorded in history, and the Chinese appear to have been the first to master the art of making and using pyrotechnics of this type several centuries back. The civilian use of fireworks for festivals and celebrations is on the increase all over the world (Figure 1.6) In India alone, several thousands of tons of fireworks and crackers are consumed on the day of Diwali (the light festival of India), submerging the entire country under a thick layer of pollutant gases of sulfur dioxide (SO_2), CO, CO_2, unburnt particulate suspension, etc., apart from the added effects of noise pollution.

Pyrotechnics have a multifarious and often a very vital role to play in military applications. Pyrotechnics are used to produce effects other than those produced by high explosives, initiating mixtures, and propellants. They produce light for illuminating or signaling purposes, heat required for an incendiary effect, smoke for signaling and

Table 1.1: General composition of propellants.

Sl. No	Type of Propellant	Composition: Major Ingredients	Composition: Minor Ingredients	Method of Preparation	Main Application
1.	Single base	NC	Plasticizers, stabilizers, flash reducers	Extrusion	Small-arms and gun ammunition
2.	Double base	NC, NG	Plasticizers, stabilizers (for rocket propellants, ballistic modifiers are also used)	Extrusion (casting for certain rocket propellants)	Gun ammunition, rockets, and missiles
3.	Triple base	NC, NG, nitroguanidine	Plasticizers, stabilizers, flash reducers	Extrusion	Large-caliber naval guns, artillery guns
4.	Nitramine base	NC, NG, RDX	Plasticizers, stabilizers (for rocket propellants, ballistic modifiers are also used)	Extrusion	Gun ammunition, rockets, and missiles
5.	Composite propellants	AP, aluminum, polymeric binder cum fuel	Plasticizers, burn rate catalysts, etc.	Casting	Rockets and missiles
6.	Composite modified double-base propellants	NC, NG, AP, aluminum	Plasticizers, burn rate catalysts, etc.	Casting	Rockets and missiles
7.	Liquid propellants for rockets	Liquid oxidizers (e.g., red fuming HNO_3, hydrogen peroxide), liquid fuels (e.g., aniline, hydrocarbons)	–	Prepared oxidizer and fuels kept in separate tanks	Rockets and missiles

screening, and intervals of time between explosive events ("delay"). Chemically, they are an intimate and finely powdered mixture of an oxidizer, a fuel (metallic or organic), and other ingredients needed for specific use such as binder, color-giving metals, etc. They are made into pellets or granules of different shapes and sizes to achieve certain parameters such as burning rate, and the making of pyrotechnics is as much an art as it is a science.

The main feature of most of the pyrotechnic reactions are (1) they are basically solid–solid reactions (in which the particle size of the reacting chemicals plays a vital role), (2) they evolve a large amount of heat in many cases, and (3) most of them hardly evolve any gas. Some of the applications of pyrotechnics for military use are as follows:

1. Producing color signals (e.g., by use of the salts of Ba, Sr, and Na for producing green, red, and yellow colors, respectively).

Figure 1.6
New Year Eve fireworks display at London Eye. *(With permission from Martin Coffin, Phoenix Fireworks Inc., Kent, UK.)*

2. Introducing a controlled or predetermined time delay in certain operations (e.g., a few milliseconds or even a few seconds delay in the operation of a fuse or explosive mixture of gases). Delay compositions are mixtures of materials that, when pressed into "delay tubes," react without evolution of gases to ensure minimum variation in the delay period (The evolution of the gases is undesirable because the pressure developed by them inside of the tube will change the reaction/burning rate and in some cases the system is not designed to withstand the pressure). Some examples of such a mixture are ($BaCrO_4 + B$) and ($KMnO_4 + Sb$)
3. Producing flares attached to an antiaircraft missile. The flares help the missile to home on to the target (e.g., $Mg + NaNO_3 +$ laminac).
4. Creating smoke for the purpose of obscuration, (e.g., $Zn + KClO_4 +$ hexachlorobenzene).

On the basis of the special effects produced by pyrotechnics, they can be categorized into four groups (Figure 1.7) as presented in the following subsections.

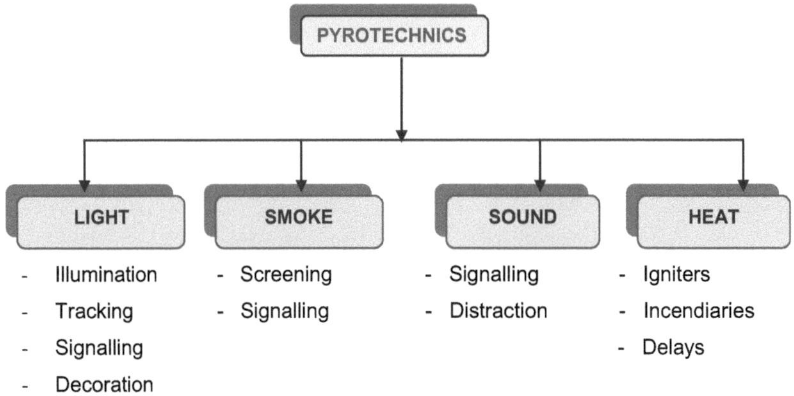

Figure 1.7
Classification of Pyrotechnics Based on Special Effects.

1.6.1 Light

Emission of bright light is the primary function of many pyrotechnic compositions. Aluminum or magnesium fuels are found in most white-light pyrotechnic compositions. These metals evolve substantial heat during oxidation and the magnesium oxide (MgO) and aluminum oxide (Al_2O_3) reaction products are good light emitters at the high reaction temperatures.

1.6.2 Smoke

Smokes are used for military signaling and screening. These are usually prepared by mixing certain dye stuffs with the fireworks. Military smokes were evolved from the mixtures of metal powders with halogenated organic compounds patented in 1920 by Captain Henri Berger of the French army. In modern warfare, special chemicals are being developed to produce smokes that stop the penetration of infrared radiation used by the enemy for detection purposes.

1.6.3 Sound

The acoustic sound wave produced by pyrotechnics is by a sudden release of high-pressure gas. Such pyrotechnics are used in various simulation devices.

1.6.4 Heat

Heat is often considered as one of the byproducts of pyrotechnics, and in some pyrotechnic applications heat or flame is the desired product. This effect can be used for either constructive or destructive purposes. Military pyrotechnic compositions as heat producers are mainly used in igniters, incendiaries, and delays.

Pyrotechnics are not conventionally classified under explosives. Each ingredient taken separately from a pyrotechnic composition may be inert. However, once they are mixed as an "oxidizer-fuel" mixture, which is also in the form of fine powder, the composition becomes an energetic material. Many of them are sensitive to impact, friction, and static electricity. Accidental initiation of pyrotechnics during a large-scale manufacture may result in the evolution of enormous heat/fire followed by disastrous detonations (some sensitive pyrotechnic dust can be initiated by as little energy as 10 μJ). Elaborate safety measures are called for while processing such compositions, such as the use of static discharge systems and nonsparking tools, maintaining at least 60% relative humidity in process buildings, mixing in liquid media in some cases, etc. The fundamental concepts of pyrotechnics are discussed in Chapter 7 of this book.

Names of international journals with their corporate office addresses, which are publishing recent research works, review papers, and new books related to energetic materials, are given in Appendix B. International societies, organizations, laboratories working in the field of explosives, propellants, and pyrotechnics are listed in Appendix C.

Appendix A

Important milestones in the development of explosives and propellants

Milestone	Year
Chinese invented black powder	~220 BC
English monk Roger Bacon experimented with black powder	1249
German monk Berthold Schwartz studied black powder	1320
Edward Howard of England rediscovered mercury fulminate	1800
Italian professor Ascanio Sobrero invented NG	1846
Schonbein at Base1 and Bottger at Frankfurt-am-Main invented NC	1845–47
Swedish scientist Immanuel Alfred Nobel set up manufacturing plant for NG	1863
Nobel's factory was destroyed	1864
Patent for Ghur dynamite	1867
Ballistite (smokeless powder)	1888
Cordite	1889
PETN (pentaerythritol tetranitrate)	1894
RDX	1899
HMX (high melting explosive)	1930
First PBX (plastic bonded explosive) composition	1952
Octol	1952
CL-20 (China Lake-20)	1987
ONC (octanitrocubane)	1999
N_5^+	2001

Appendix B

International journals in the field of HEMs

Name of the Journal	Address
Propellants, Explosives, Pyrotechnics	Journal Customer Services, Wiley, 350 Main Street, Malden, MA 02148, USA Phone: 1-781-388-8598 or +1-800-835-6770; E-mail: cs-journals@wiley.com
Journal of Energetic Materials	subscriptions@tandf.co.uk
Journal of Pyrotechnics	Bonnie Kosanke, 1775 Blair Road, Whitewater, CO 81527, USA. Phone: 1-970-245-0692; Fax: 1-970-245-0692; E-mail: bonnie@jpyro.com
Combustion and Flame	Dan O'Connell, Publicity Manager, Science & Technology Books Phone: 1-781-313-4726
Defense Science Journal	Director, DESIDOC, DRDO Metcalfe House, Delhi- 110 054 India. E-mail: dsj@desidoc.drdo.in
Science and Technology of Energetic Materials	Japan Explosives Society, Kaseihin Kaikan Building. 5-18-17, Roppongi, Minato-ku, Tokyo 106-0032, Japan. Phone: +81-3-5575-6605; Fax: +81-3-5575-6607; E-mail: web-master@jes.or.jp

Appendix C

International societies working in the field of explosives, propellants, and pyrotechnics

Name of the Society	Address
Institute of Chemical Technology, Germany	Dr. Stefan Tröster, Fraunhofer-Institut für Chemische Technologie ICT, Joseph-von-Fraunhofer-Straße 7, 76327 Pfinztal, Germany. Phone: +49-721-4640-392
Institute of Detonation	Christopher Boswell, IHDIV, NSWC. Phone: 1-301-744-4619; E-mail: intdetsymp@navy.mil

International societies working in the field of explosives, propellants, and pyrotechnics—cont'd

Name of the Society	Address
American Institute of Aeronautics and Astronautics (AIAA)	AIAA Headquarters, 1801 Alexander Bell Drive, Suite 500, Reston, VA 20191-4344 USA. Phone: 1-703-264-7500 or 1-800-639-AIAA; Fax: 1-703-264-7551
Japan Society of Energetic Materials	Japan Explosives Society, Ichijoji Building, 3F, 2-3-22, Azabudai, Minato-ku, Tokyo 106-0041, Japan. Phone: +81-3-5575-6605; Fax: +81-3-5575-6607; E-mail: webmaster@jes.or.jp
High Energy Materials Society of India	High Energy Materials Research Laboratory (HEMRL), Sutarwadi, Pune-411021. Fax: 020-25869697; Website: www.hemsichd.org; E-mail: hemce2011@gmail.com

Suggested Reading

[1] S.M. Kaye (Ed.), Encyclopedia of Explosives and Related Items, vol. 1–10, U. S. Army Armament R&D Command, N.J, 1983.
(NOTE: This is the most exhaustive compilation carried out on explosives and related items. To be used for "reference" and not "reading.")

[2] R. Meyer, J. Kohler, Explosives, VCH Publishers, Germany, 1993 (Encyclopedia – handy for referencing).

[3] T. Urbanski, Chemistry and Technology of Explosives, vol. 1–4, Pergamon Press, Oxford, New York, 1983.
(Considered to be the Bible of explosives chemists and technologists – a 'must' reference book in any lab/institution/factory dealing with high energy materials.)

[4] Service Text Book of Explosives, Min. of Defence, Publication, UK, 1972.

[5] B. Morgan, Explosions and Explosives, Macmillan (Quantum Books), London, New York, 1967.

[6] A. Bailey, S.G. Murray, Explosives, Propellants and Pyrotechnics, Pergamon Press, Oxford, New York, 1988.

[7] T.L. Davis, The Chemistry of Powder and Explosives, Wiley, New York, 1956.

Questions

1. Who were the first to invent gunpowder? And, who was the first scientist to describe its chemical constituents?
2. What are the roles played by sulfur, KNO_3, and charcoal in gunpowder?
3. Write the chemical equation for the explosive reaction of gunpowder?

4. Write the molecular structure of TNT and indicate the fuel, oxidizer, and energetic moieties in it.
5. How could Alfred Nobel tame the dangerous NG? What do you think is the mechanism behind it?
6. Define explosives.
7. Classify the following explosives as primary, secondary, or low explosives: (a) Lead azide, (b) β-HMX, (c) TNT, (d) a rocket propellant, (e) tetrazene, (f) PETN, (g) RDX, (h) a gun propellant, (I) tetryl, (j) mercury fulminate.
8. Why is a molecular explosive preferred to a mixture that acts as an explosive?
9. What important characteristics should a military explosive satisfy?
10. Why can you not use a military explosive for civilian application (and vice versa)?
11. What are explosophores?
12. How would you classify propellants with respect to (a) application and (b) composition.
13. What are the different applications of pyrotechnics?
14. How do the following differ from each other? (a) a fuel, (b) a propellant, (c) a high explosive, and (d) a pyrotechnic.
15. What are the major and minor ingredients of common double- and triple-base propellants?
16. Name some applications of pyrotechnics in the military.

CHAPTER 2

Energetics of Energetic Materials

2.1 Are Explosives and Propellants High-Energy Materials?

Explosives are storehouses of energy. The potential energy lying dormant in the molecules of these materials is released when they are properly triggered or initiated and the release of this energy originates at the breaking of the energetic chemical bonds in the molecule of the explosive compound. The explosives (and propellants, which are categorized as low explosives) are sometimes referred to as "high-energy materials" (HEMs) or "energetic materials." Is one justified in using these terms for explosives? Let us compare the most powerful high explosive in use today—HMX (an abbreviation for High Melting Explosive; chemical name being cyclic tetramethylene tetranitramine)—with a well-known fuel—coal—in terms of their energetics.

From Table 2.1, it is seen that for every gram, coal produces more than 5 times the heat produced by HMX. The heat evolved by 1 g of coal and HMX is illustrated as a chart in Figure 2.1. A similar comparison will show us that all fuels of day-to-day use release far more heat than any known explosive. Therefore, in a thermochemical sense, explosives and propellants are not really HEMs. However, we also observe that HMX detonates in much less time than a piece of coal takes to burn. The former undergoes the process of detonation accompanied by shock waves whereas the latter takes its own time for combustion with the help of oxygen available in air. If we take the rate at which the heat is released, then the power of HMX is approximately 5.6×10^9 W in comparison to 488 W of coal in the above example. This power generation by HMX is far more than the capacity of all of the power generators in the country put together. The better term for explosives may not be HEMs but "power-packed materials."

In the above example, an arbitrary figure of 60 s has been given for the burning of coal. Under static wind conditions, the time taken for 1 g of coal to burn depends on its surface area exposed to air. As we go on breaking it into pieces, the burning time of coal comes

Table 2.1: Heat generated by coal and HMX.

Sl. No.	Property	Coal (1 g)	HMX (1 g)
1.	Heat evolved	7000 cal (heat of combustion)	1355 cal (heat of explosion)
2.	Time (burning/detonation)	60 s	10^{-6} s
3.	Power	488 W	5.6×10^9 W

Demystifying Explosives: Concepts in High Energy Materials. http://dx.doi.org/10.1016/B978-0-12-801576-6.00002-1

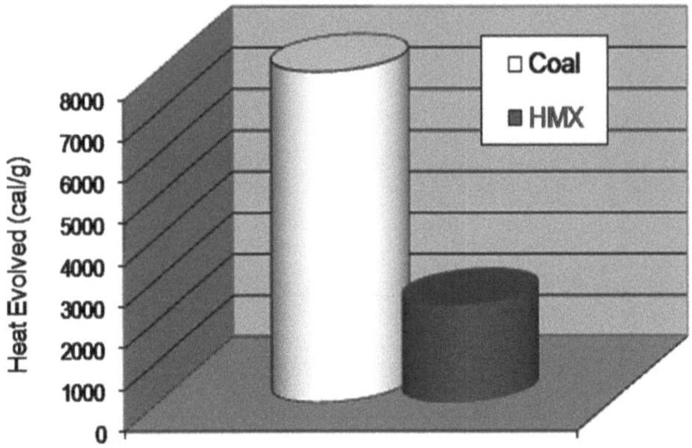

Figure 2.1
Comparison of Heat Evolved by 1 g of Coal and HMX.

down drastically. At its extreme, when the same 1 g of coal is finely powdered and dispersed as coal dust in air, facilitating the exposure of the maximum surface area to air, each such dust particle is in intimate contact with the oxygen molecules of air. When initiated, the combustion reaction takes place so fast that it is virtually converted into a violent detonation. Disastrous coal-dust explosions in coal mines are a result of this phenomenon. Such dust explosions are not uncommon in many other industries.

2.2 Explosive: The Wonderful Lamp

An explosive is similar to the genie that we come across in the ever-fascinating tale of Aladdin and the Wonderful Lamp. It has great potential, but then it has to be kept under check or "bottled-up." Only when its services are needed do we open the bottle, and, in the case of explosives, we give the necessary trigger energy. An explosive is a substance in a metastable equilibrium, in a "ready-to-go-off" stage with huge potential energy. The relation between the energy needed to make an explosive and the energy released by it on explosion can be qualitatively understood by comparing it with a huge boulder brought to the apex of a cliff.

Figure 2.2(a) and (b), respectively give an analogy between a boulder kept on the brink of a cliff and an explosive synthesized and "kept" in a metastable state. One has to make great efforts (or spend much energy) to place the boulder on the cliff (A) in Figure 2.2(a). The boulder continues to remain there until someone decides to push it (giving an energy equal to B) so that it falls off from a great height, converting the potential energy into kinetic energy, which is dissipated as heat and sound when it strikes the ground. Release of energy is equal to C. Likewise, the synthesis of an explosive molecule is done by packing in it a great amount of potential energy such as high bond energy, structural strain, etc., and it is kept in the metastable state as shown in Figure 2.2(b). (D−E) is the

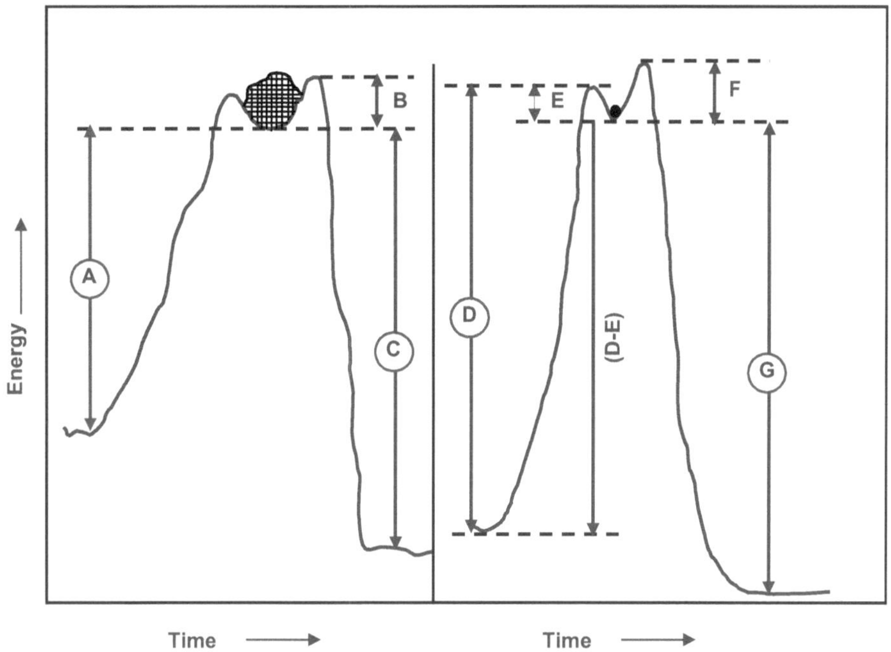

Figure 2.2
A Boulder on a Cliff and an Explosive Molecule.

effective energy spent in such a synthesis. If the reactants are assumed as elements such as carbon, hydrogen, and oxygen, (D−E) is referred to as the heat of formation of the explosive. The explosive now needs only a trigger energy (generally called activation energy) equal to F, so that a net energy equal to G is liberated during the explosive process and the formation of stable products.

The chemist who wants to synthesize an explosive ensures that (1) as far as possible the product has a high positive heat of formation (i.e., the energy level of the explosive molecule is higher than that of the elements from which it is made), (2) it has its own supply of oxygen in the molecule to be independent of external or atmospheric oxygen to affect the process of explosion, and (3) the explosive reaction results in a large amount of gases.

Factors 1 and 2 will ensure that the explosion process releases a large amount of heat (heat of explosion), thereby enormously increasing the temperature of the products, normally more than 2000 °C. Factor 3 will ensure that, with so many gases at a high temperature, there will be development of very high pressures. The gases expand rapidly from very high pressures to the atmospheric pressure, thereby performing a large amount of work in a short time; that is, the produced gases will work as a powerful working fluid to perform certain assigned tasks such as the blast effect produced by high explosives in microseconds, the work of throwing a projectile through a gun barrel in a few milliseconds, or the self-propulsion by a rocket in a time period varying from a few seconds to even a few minutes.

Is an oxidation reaction always necessary in a chemical explosion? Although most of the chemical explosions involve fast oxidation of fuel elements, it need not be so in some cases. For example, lead azide ($Pb(N_3)_2$), a well-known primary explosive, does not contain any oxygen atoms in its molecule. However, it has a positive heat of formation. The azide ($-N-N{\equiv}N$) groups attached to the lead atom have weak linkages and are themselves at a higher energy level. Only a small trigger energy is necessary to rupture these linkages to produce more stable products with the evolution of energy.

$$Pb\left(-\overset{(-)}{N}-\overset{(+)}{N}\equiv N\right)_2 \rightarrow Pb + 3N_2 + 110.8\ \text{kcal}$$

2.3 Thermochemistry and Explosive Energy

Chemical reactions are accompanied by energy changes, mainly in the form of heat. The branch of science that deals with the heat changes during chemical reactions is called "thermochemistry." It is essential to remember certain basic concepts in thermochemistry to obtain better insight into the heat transactions during the formation and explosion of explosives. The concepts about the three important parameters—internal energy (E), heat content or enthalpy (H), and work (W)—should also be clear. The internal energy of a substance is the total quantity of energy it possesses by virtue of its kinetic portion of energy (due to translational, vibrational, and rotational motions associated with the molecules) and the potential portion of energy (due to various interatomic, intermolecular, and submolecular forces of attraction and repulsion). In a chemical reaction in which certain bonds of the reactant molecules are broken and certain bonds of the product molecules are formed, it is mostly the kinetic portion of the internal energy that undergoes a change and may be positive or negative. E is a measure of the absolute temperature of the given substance. When the temperature increases, all of these kinetic aspects of energy obviously increase in the molecule.

Heat content, or enthalpy (H), which is defined as $H = E + PV$, where P and V are pressure and volume, respectively, comes in picture when the system undergoes a change in its E and, at the same time, a certain amount of work is also performed. The work, W, is taken as $W = P(\Delta V)$, where ΔV is the change in volume due to the work of expansion. The absolute values of E and H have no significance. We are interested only in the changes in the values of E and H (i.e., ΔE and ΔH, respectively) of a system when changes such as a chemical reaction take place.

2.3.1 Heat of Reaction

The net heat absorbed or evolved during a chemical reaction is known as the heat of reaction (ΔH_r). The reaction involves the expenditure of heat energy to break certain bonds in reactant molecules and the evolution of heat energy when certain bonds are made in the product molecules. If the expenditure is more than the evolution, then it is an endothermic reaction and the converse is true for an exothermic reaction. In an endothermic reaction, ΔH of the products is more than that of the reactants and the converse is true for an exothermic reaction.

2.3.1.1 An Endothermic Reaction (Net Heat Absorption)

Heat content	A	+	B	→	C	+	D	—	Heat
(arbitrary values)	80 cal		100 cal		120 cal		100 cal		40 cal

$$\begin{aligned}\Delta H &= H_{(products)} - H_{(reactants)} \\ &= (120 + 100) - (100 + 80) = 40 \text{ cal}, \quad \Delta H = +40 \text{ cal}\end{aligned} \tag{2.1}$$

In its general form, an endothermic reaction can be represented as

$$\text{Reactants} + (\text{Energy absorbed by reactants}) \rightarrow \text{Products}$$

2.3.1.2 An Exothermic Reaction (Net Heat Evolution)

$$\underset{100\text{ cal}}{P} + \underset{150\text{ cal}}{Q} \rightarrow \underset{50\text{ cal}}{R} + \underset{75\text{ cal}}{S} + \underset{125\text{ cal}}{\text{Heat}}$$

$$\Delta H = H_{(products)} - H_{(reactants)}$$
$$= (50 + 75) - (100 + 150) = -125 \text{ cal} \quad \Delta H = -125 \text{ cal}$$

In its general form, an exothermic reaction can be represented as

$$\text{Reactants} \rightarrow \text{Products} + \text{Energy released}$$

It should be remembered that ΔH is negative for exothermic reactions and positive for endothermic reactions. All explosive chemical reactions are exothermic, and the ΔH values of these reactions are negative.

All heats of reactions are compared at a standard state, which is normally defined as 25 °C (298 K) and 1 atm pressure.

2.3.2 Heat of Formation

The heat of formation (ΔH_f) of a compound is defined as "the heat evolved or absorbed when 1 mole of the compound is formed from its elements in the standard state."

It is assumed that the heat of formation of elements (e.g., H, O, C, etc.) is taken as zero. The heats of formation of explosives can be either exothermic or endothermic; for nitroglycerine (NG), its heat of formation reaction is written as

$$3C + 5/2H_2 + 3/2N_2 + 9/2O_2 \rightarrow C_3H_5(NO_3)_3 + 84\,\text{kcal}(\Delta H_f = -84\,\text{kcal/mol})$$

$Pb(N_3)_2$, a primary explosive, is formed endothermically.

$$Pb + 3N_2 \rightarrow Pb(N_3)_2 - 110.8\,\text{kcal.}$$
$$(\Delta H_f = +110.8\,\text{kcal/mol})$$

As we will see shortly, ΔH_f is a very important thermochemical parameter for an explosive because it plays a vital role in the heat of explosion or heat of detonation and other related parameters. An explosive that has a positive heat of formation or a low negative value of heat of formation is preferred for release of maximum heat during its explosion.

2.3.2.1 Experimental Estimation of ΔH_f

In most cases, it is not possible to experimentally form a compound from its elements. For example, we cannot synthesize NG from its elements of C, H, N, and O as shown above. It is not that simple to produce NG. We use Hess's law of constant heat summation to solve this problem. It states: "If a chemical reaction is carried out in stages, the algebraic sum of the amounts of heat evolved in separate stages is equal to the total amount of heat evolved when the reaction occurs directly." That is, the resultant heat change in a chemical reaction performed either at constant pressure or constant volume is the same whether it takes place in one or several stages. Thus, according to this law, ΔE and ΔH are dependent only on the initial and final states and not on the path connecting them. It is pictorially represented in Figure 2.3. This law is a corollary of the Law of Conservation of Energy.

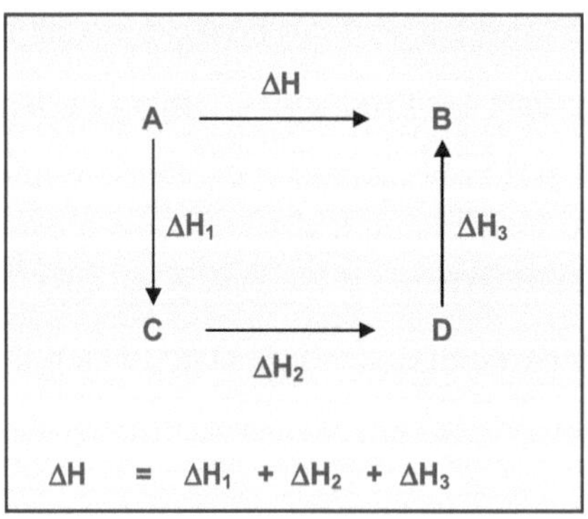

Figure 2.3
Hess's Law.

Another way of stating Hess's law is

Heat of the reaction (x) = $\sum(\Delta H_f)$ products − $\sum(\Delta H_f)$ reactants, or

$$\text{i.e.,}\quad x = (\Delta H_f)_C + (\Delta H_f)_D - (\Delta H_f)_A + (\Delta H_f)_B \qquad (2.2)$$

where A and B are the reactants and C and D are the products of a reaction.

Coming back to the example of NG, how can we calculate its ΔH_f? (Data given are heat of explosion of NG, $\Delta H_e = -367$ kcal/mol; $\Delta H_{f(CO_2)} = -94$ kcal/mol; and $\Delta H_{f(H_2O)} = -67.4$ kcal/mol.)

The explosive reaction of NG can be written as

$$C_3H_5(NO_3)_3 \rightarrow 3CO_2 + 5/2H_2O + 3/2N_2 + 1/4O_2 + 367\ \text{kcal}$$

(Note: The heat of explosion can be experimentally determined.)

Using Hess Law, $\Delta H_{reaction} = \sum(\Delta H_f)_{prod} - \sum(\Delta H_f)_{react}$

$$-367 = [(3 \times -94) + (5/2 \times -67.4)] - [(\Delta H_f)_{NG}]$$

(Note: ΔH_f of the elements is taken as zero.)

$$\Delta H_f \text{ of NG} = -83.5\ \text{kcal/mol}$$

Therefore, NG is an exothermic compound. Alternatively, if the ΔH_f value of an explosive is available, then we can calculate its heat of explosion.

2.3.2.2 Theoretical Prediction of ΔH_f

There are many computer codes to theoretically evaluate or predict the performance of high explosives (e.g., TIGER, BKW Code), rocket propellants (e.g., NASA-LEWIS), and gun propellants (e.g., BLAKE). However, none of these can operate without having the data on ΔH_f of the concerned energetic materials and their products of explosion. There are several potential candidate molecules of explosives that are yet to be synthesized. In addition, if we want to theoretically predict their performance as high explosives or propellant ingredients (and decide whether it is worth synthesizing them), then we need to know their ΔH_f values. Naturally, we cannot resort to the experimental method (as in the case of NG) because the compound is still not available. This section briefly mentions a few theoretical methods available for the purpose of predicting the ΔH_f values.

1. Group additivity method:
   In this method, the explosive molecule is divided into different groups and each group is assigned an enthalpy value. The individual group enthalpies are added to give the ΔH_f of the molecule. This method neglects the effect of intergroup interactive forces. It

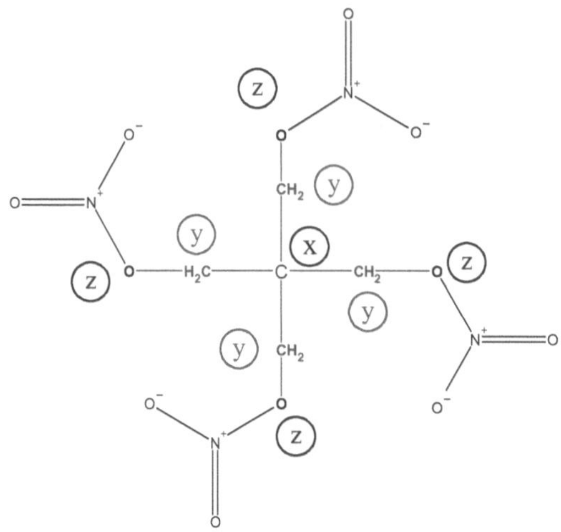

Figure 2.4
Pentaerythritol Tetranitrate Molecule.

is mainly applicable to gases, and for solids a correction by approximately 25 kcal/mol is applied as heat of sublimation. Taking the example of the explosive pentaerythritol tetranitrate (PETN; Figures 2.4 and 2.5), we see that in a PETN molecule the following groups are available:

a. One $C-(C)_4$ group marked by $\Delta H_f = +0.50$ kcal/mol (i.e., central carbon atom)
b. Four $C-(CH_2)-(O)$ groups marked by $\Delta H_f = -8.1$ kcal/mol
c. Four $C-(O-NO_2)$ groups marked by $\Delta H_f = -19.4$ kcal/mol

$$\begin{aligned}\Delta H_f\,(\text{PETN}) &= (1 \times 0.5) + (4 \times -8.1) + (4 \times -19.4)\\ &= -109.5\ \text{kcal/mol}\end{aligned}$$

Subtracting the assumed heat of sublimation,

$$\begin{aligned}\Delta H_f(\text{PETN}) &= -109.5\ \text{kcal/mol} - 25\ \text{kcal/mol}\\ &= -134.5\ \text{kcal/mol} \quad (\text{Experimental value} = -128.7\ \text{kcal/mol})\end{aligned}$$

S. W. Benson has determined the ΔH_f values for many groups in aliphatic, aromatic, and heterocyclic compounds. These data are of great help in calculating, at least approximately, the ΔH_f values of many compounds.

2. Other methods:
The Russian scientist Dmitrii V. Sukhachev and colleagues have recently evolved a method based on Quantitative Structure—Property Relationship and Efficient Modelling of Molecular Activity software to estimate and predict the ΔH_f values of nonaromatic polynitro compounds. This approach is based on the construction of regression equations that relate the structure of known compounds to their physical, chemical, and

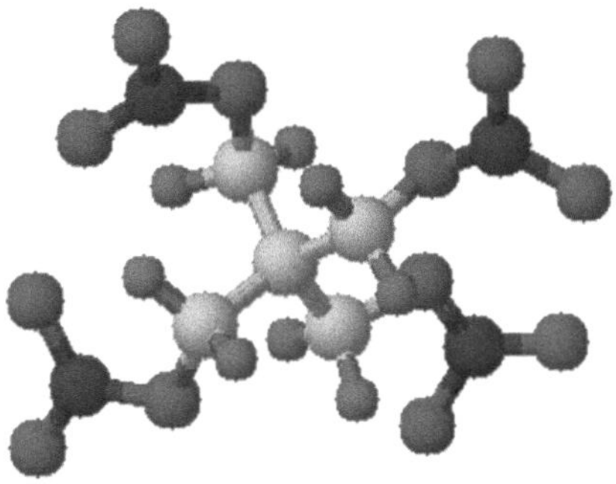

Figure 2.5
Ball and Stick Model of Pentaerythritol Tetranitrate Molecule.

topological properties. The best model equation is chosen and is used to predict the properties of novel structures and select potentially active structures for further synthesis.

Software packages that are based on a quantum mechanical approach have been developed to predict the ΔH_f values with better accuracy.

2.3.3 Heat of Explosion (ΔH_e) and Heat of Combustion (ΔH_c)

Most of the explosives contain C, H, N, and O atoms. During an explosive reaction, the molecule uses the oxygen atoms available within it and does not depend on the external, atmospheric oxygen; it probably has no time for that because of the fast nature of the explosive process. It must be noted that all oxidation reactions ($C \rightarrow CO$, $CO \rightarrow CO_2$, $H \rightarrow H_2O$) and all explosive reactions are exothermic. Let us now distinguish between two types of oxidation reactions—combustion and explosion.

1. Heat of combustion (ΔH_c): This is defined as the heat evolved when 1 mole of a compound is completely burnt in excess of oxygen. It means that all C atoms and H atoms in the molecule are converted into carbon dioxide (CO_2) and water (H_2O), respectively. A fuel burning in air gives out heat of combustion. Heat of combustion is often referred to as the "calorific value." The amount of heat produced in our body when certain components of food such as fat undergo combustion during metabolism is referred to by this term or simply "calories," a term of which we are so conscious about today.
2. Heat of explosion (ΔH_e): If we take explosives, barring a few examples such as NG, we find that the amount of oxygen available in their molecules is not sufficient to convert C and H atoms completely to CO_2 and H_2O, respectively. Then it becomes a competition between the C and H atoms to get themselves oxidized. However, the end result is that the products of explosion are underoxidized, containing carbon monoxide (CO) and

sometimes H_2 and C also. It is evident that the heat of explosion is always less than the heat of combustion. The underoxidized products of explosion are themselves fuels, and, in cases of explosion, one finds the secondary fire balls formed due to the further oxidation of these underoxidized products by atmospheric oxygen.

Conventionally, the term heat of explosion is applied to propellants (which are deflagrating "explosives") and heat of detonation is used for high explosives. Dunkel defines the heat of detonation as the heat evolved when an explosive detonates and the products are still at the Chapman–Jouguet condition (see Chapter 3), in which the gases are at approximately 5000 K and 10^5 atm in the detonation zone. The composition of the products at the detonation zone is slightly different from what we may find in a calorimeter; therefore, the heat of detonation as defined by Dunkel will be slightly different from the heat of explosion. For all practical purposes, we can take that the heat of detonation is almost synonymous with the heat of explosion. We will use the term heat of explosion for all purposes of calculation. Heat of explosion (ΔH_e) is also referred to as the calorimetric value (shortly "cal.val"). The explosives and propellants depend only on the oxygen available in their molecules; therefore, they can function even in vacuum. Moreover, the explosive reactions are too fast to rope in the atmospheric oxygen even if it is available. In view of this, cal.val (ΔH_e) plays a very important role in the field of explosives and we rarely talk about calorific value (ΔH_c).

2.3.3.1 Need to Standardize Calorimetric Value

ΔH_e and ΔH_c are experimentally determined using a bomb calorimeter, the experimental details of which can be had from any standard book on explosives and propellants. In the case of determination of cal.val, a fixed weight of the explosive (or propellant), usually approximately 2.5 g, is ignited and exploded after purging the bomb calorimeter with nitrogen or helium to ensure that no oxygen due to residual air is present at the time of explosion. The heat evolved is calculated by measuring the increase in temperature of water in the calorimeter. The cal.val obtained is for the reaction in which water is obtained in the liquid form. In the case of ΔH_c (calorific value) determination, to ensure complete combustion, the calorimeter is flushed with an excess of oxygen before the ignition of the explosive.

For a given explosive compound, ΔH_c is a standard value whereas ΔH_e is not. There is a need to standardize the conditions under which ΔH_e is to be measured. Let us assume that in the first experiment we take 2.5 g of an explosive and perform its ΔH_e determination in a bomb calorimeter, the volume of which is 700 cc (i.e., loading density of the explosive = 2.5/700 g/cc) and let the cal.val obtained be Q_1 cal/g. If the experiment is repeated with, say, 5 g of the same explosive in the same calorimeter (loading density = 5/700 g/cc, i.e., twice as in the case of first experiment), then the cal.val obtained will be different, say Q_2 cal/g. In the second experiment, after the explosion, the pressure of the product gases will be higher in comparison to the first experiment because of increased loading density. At higher pressures, the product gases (a mixture of CO, CO_2, H_2O, probably some H_2 and C also in the case of explosives with low oxygen balance (OB)) undergo a shift in the equilibrium and the resultant heat output will be different. Therefore, cal.val experiments should be conducted under standardized conditions, particularly with respect to loading density. (In the case of ΔH_c determination, this problem does not arise because all of the products are already in a completely oxidized condition).

2.3.3.2 Partial Heat of Explosion

Schmidt proposed a simplified way of estimating the probable heat of explosion of a propellant containing explosive and nonexplosive ingredients. In this method, a partial heat of explosion is assigned to each component and materials with high negative OB (e.g., stabilizers, gelatinizers) are assigned negative values. The heat of explosion of the propellant is calculated by the addition of the partial values weighted in proportion to the respective percentage of the individual components.

It is interesting to note that in the case of compositions containing NG (which has a positive OB), the heat of explosion value is more than the calculated value because the excess oxygen from NG reacts with the carbon of other components to produce more heat.

2.3.4 Oxygen Balance

The percentage excess or deficit of oxygen present in a compound required for its complete oxidation to CO_2, H_2O, etc., is known as the OB of that compound.

If the compound has less oxygen in its molecule than that required for complete oxidation, then it is said to have a negative OB and vice versa.

Example 1: NG (Figure 2.6) has a positive OB. Its explosive reaction can be written as

$$C_3H_5(NO_3)_3 \rightarrow 3CO_2 + 5/2H_2O + 3/2N_2 + 1/4O_2(+\text{heat})$$
$$(\text{Mol.Wt.} = 227.1)$$

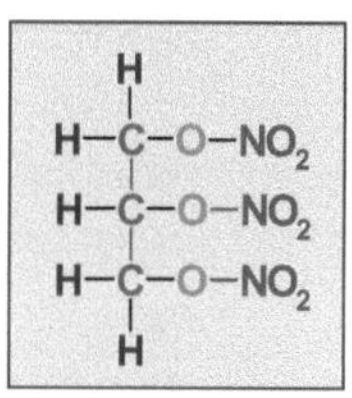

Figure 2.6
Nitroglycerine.

We find that 227.1 g of NG (1 mole weight of NG) has enough oxygen in its molecule so as to evolve an excess of oxygen ($1/4 O_2 = 8$ g of oxygen) even after completely oxidizing C and H to CO_2 and H_2O, respectively.

$$\begin{aligned} &\text{227.1 g of NG evolves} && 8 \text{ g of } O_2 \\ &\text{Therefore 100 g of NG evolves} && 8/227.1 \times 100 \text{ g of } O_2 \\ & && = 3.5\% \\ &\text{OB of NG} && = +3.5\% \end{aligned}$$

Example 2: Trinitrotoluene (TNT; Figure 2.7) has a negative OB. In the TNT molecule ($C_7H_5N_3O_6$), we can see that the number of oxygen atoms (6) is very insufficient to completely oxidize 7 carbon atoms (14 oxygen atoms needed for the $7C \rightarrow 7CO_2$ oxidation) and 5 hydrogen atoms (5/2 oxygen atoms needed for the $5H \rightarrow 5/2H_2O$). Compared with 14 + 5/2 (i.e., 33/2 oxygen atoms), TNT has only 6 oxygen atoms to achieve complete oxidation. This deficit (i.e., 33/2 vs 6; i.e., 21/2 oxygen atoms, 21/4 oxygen molecules) must be written on the left-hand side of the TNT combustion equation as follows:

$$C_7H_5N_3O_6 + 21/4\, O_2 \rightarrow 7CO_2 + 5/2\, H_2O + 3/2\, N_2$$
$$(\text{Mol.Wt} = 227.1)$$

Therefore, 227.1 g of TNT requires 168 g of oxygen (corresponds to 21/2 oxygen atoms), and 100 g of TNT requires $168/227.1 \times 100 = 74$ g of oxygen. Therefore, the OB of TNT $= -74\%$.

For a CHNO explosive with the formula $C_xH_yN_wO_z$, the OB percentage can be found by a general formula as follows:

$$\text{OB\%} = \frac{100 \times \text{At.Wt.of oxygen}}{\text{Mol.Wt. of the compound}}\left(Z - 2x - \frac{y}{2}\right)$$

OB is one of the important parameters of HEMs. The ideal OB of an explosive compound is zero. When it has a negative OB, the products of explosion contain underoxidized CO and there may also be some H_2. It means that if there had been some more oxygen, we

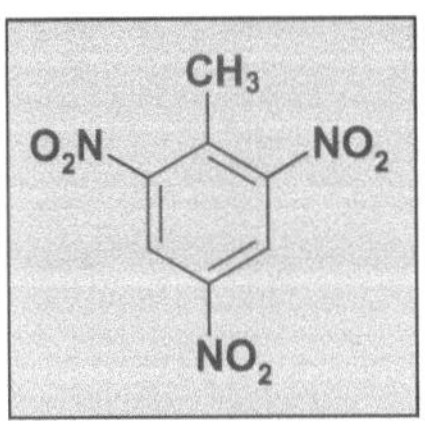

Figure 2.7
Trinitrotoluene.

could have got more heat by further oxidation of CO and H_2 to CO_2 and H_2O, respectively. In the case of compounds with positive OB, the extra oxygen evolved after complete oxidation reactions (e.g., NG) does not serve us any purpose. It has been there in the molecule as a sort of "dead weight."

Figure 2.8 shows that the ideal OB for an explosive is zero when maximum heat can be obtained by an explosive reaction from a given weight of explosive. However, from the OB values of several explosives, it is seen that, known explosives, barring NG, have negative OB values (e.g., nitrocellulose (NC) in the range of approximately −28%, TNT = −74%, research and development explosive (RDX) = −21.6%). Therefore, it is not possible to formulate a military explosive or propellant composition with zero OB. Most of these compositions have negative OB values. As we will see in subsequent sections, a factor that is as important as heat output is the number of moles of gases evolved from a unit weight of the explosive or propellant (n). The higher the value of n in the explosion/deflagration products, the higher will be the performance of the HEM. Naturally, for a gram of explosive or propellant, a higher value of n means a lower value for the average molecular weight (M) of gaseous products. The value of n plays a vital role in the field of explosives and propellants. Thus, if we get smaller molecules such as CO and H_2 instead of CO_2 and H_2O, we must remember that what we lost as heat output is compensated for, at least partly, by what we gain as work output due to higher values of n.

On the other hand, OB in commercial explosives cannot be very negative. They should have an OB close to zero. If it is negative, then the amount of toxic gases such as CO and in some cases nitrous oxide evolved will be unacceptable.

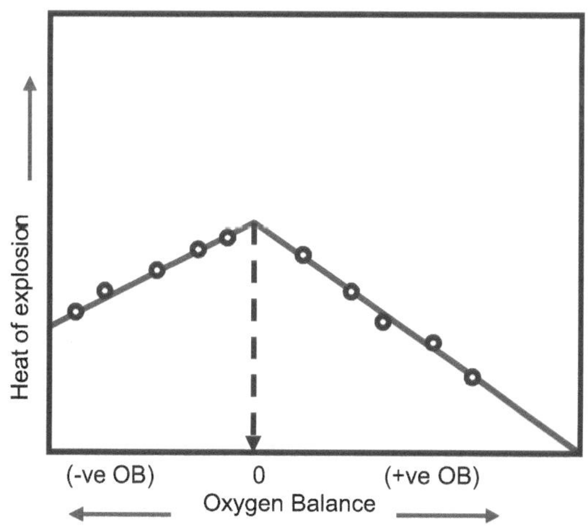

Figure 2.8
Plot of Heat of Explosion against Oxygen Balance (OB).

Worked Example 2.1

Calculate the following parameters for RDX: (1) OB, (2) heat of explosion, and (3) heat of combustion.

(Given: heats of formation of RDX, CO, CO_2, and $H_2O_{(l)}$ are +16.09, −26.7, −94.05, and −67.42 kcal/mol, respectively).

1. The molecular formula of RDX is $C_3H_6N_6O_6$, which corresponds to a molecular weight of 222.
   It needs three extra oxygen atoms to completely oxidize C and H to CO_2 and H_2O, respectively. The combustion equation is written as

$$C_3H_6N_6O_6 + 3/2O_2 \rightarrow 3CO_2 + 3H_2O + 3N_2\ (+\Delta H_c)$$

   222 g of RDX requires 48 g of oxygen.
   Therefore, 100 g of RDX requires 48/222 × 100 g oxygen = 21.6%.
   Therefore, the OB of RDX = −21.6%.
2. Heat of combustion (ΔH_c)
   From the above equation, we can write

$$\Delta H_c = \Sigma(\Delta H_f)\text{products} - \Sigma(\Delta H_f)\text{reactants}$$

$$\Delta H_c = [(3 \times -94.05) + (3 \times -67.4)] - (16.09)$$

$$= -500.5\ \text{kcal/mole} = -\frac{500{,}500\ \text{cal/g}}{222} = -2255\ \text{cal/g}$$

3. Heat of explosion (ΔH_e; in which no external oxygen participates in the reaction)
   The explosion reaction of RDX can be written as

$$C_3H_6N_6O_6 \rightarrow 3CO + 3H_2O + 3N_2\ (+\Delta H_e)$$

$$\Delta H_e = \Sigma(\Delta H_f)\text{products} - \Sigma(\Delta H_f)\text{reactants}$$

$$= [(3 \times -26.71) + (3 \times -67.4)] - (16.09)$$

$$= -298.4\ \text{kcal/mole} = -\frac{298{,}400}{222}\ \text{cal/g}$$

$$= -1344\ \text{cal/g}$$

   Compared with ΔH_e, ΔH_c is more by 68%.

2.3.5 Heat of Explosion: Dependence on Heat of Formation and Oxygen Balance

2.3.5.1 Balance

We have seen from Figure 2.8 that ΔH_e has a dependence on OB and has the maximum value at zero OB. ΔH_f values are of great importance for all HEMs. Even when new or potential compounds are to be targeted for synthesis of futuristic explosives, extensive computerized calculations are made to know their ΔH_f values. This is mostly because a positive value (or a low negative value) of ΔH_f for an explosive ensures that the explosive reaction gives out a large amount of heat.

Figure 2.9 qualitatively illustrates the effect of ΔH_f on ΔH_e. Explosive A is formed from its elements ($\Delta H_f = +x$) and later explodes to form stable products ($\Delta H_e = a$). A similar depiction is given for explosive B, which has a negative value for ΔH_f ($= -x$). Its heat of explosion ($\Delta H_e = b$) is much less than that of explosive A. Therefore, one expects higher ΔH_e values for an explosive that has a positive value of ΔH_f.

However, there can be a few exceptions. $Pb(N_3)_2$ is an endothermic compound ($\Delta H_f = +340$ cal/g), and NG is an exothermic compound ($\Delta H_f = -392$ cal/g). Their ΔH_e values are -381 and -1617 cal/g, respectively. This means that although $Pb(N_3)_2$ has a positive heat of formation, its heat output during explosion is far less than that of NG,

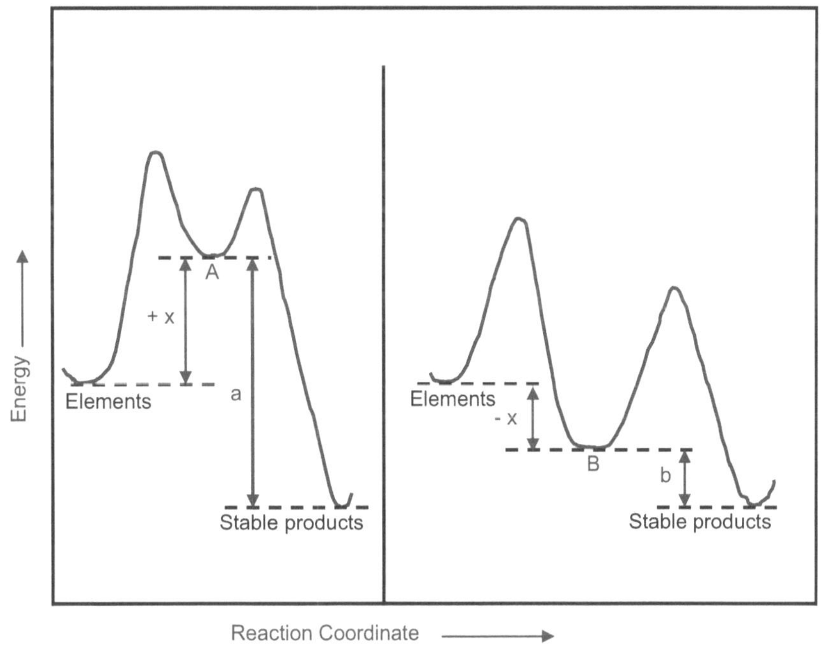

Figure 2.9
Effect of ΔH_f on ΔH_e.

which has a negative heat of formation. This is because a molecule of NG has sufficient oxygen atoms that result the in highly exothermic oxidation of C and H atoms whereas, despite its positive ΔH_f value, $Pb(N_3)_2$ is not privileged even with a single oxygen atom, and the limited heat output it gives is due to the breakage of the energetic azide linkage.

The combined effect of ΔH_f and OB on ΔH_e was studied by Edward Baroody and colleagues. Figure 2.10 shows a plot of the ΔH_f and OB of some well-known CHNO explosives. It is seen that the higher the energy output from the compound, the more it shifts toward the top right-hand corner; in the reverse case, it is toward the bottom left-hand corner.

2.3.6 OB of Composite Explosives

Mostly, for military and industrial purposes, mixtures of different explosives and other chemicals are used rather than a single explosive. These are termed "composite explosives." A common example is composition B-3, which is made up of a 64/36 mixture of RDX and TNT. It can be calculated that this composition would have an OB of −40.5%. Taking another example, ANFO is a simple mixture of prilled ammonium nitrate (AN) and fuel oil (FO) at a nearly zero oxygen balanced ratio of 94/6 AN/FO. The 6% oil

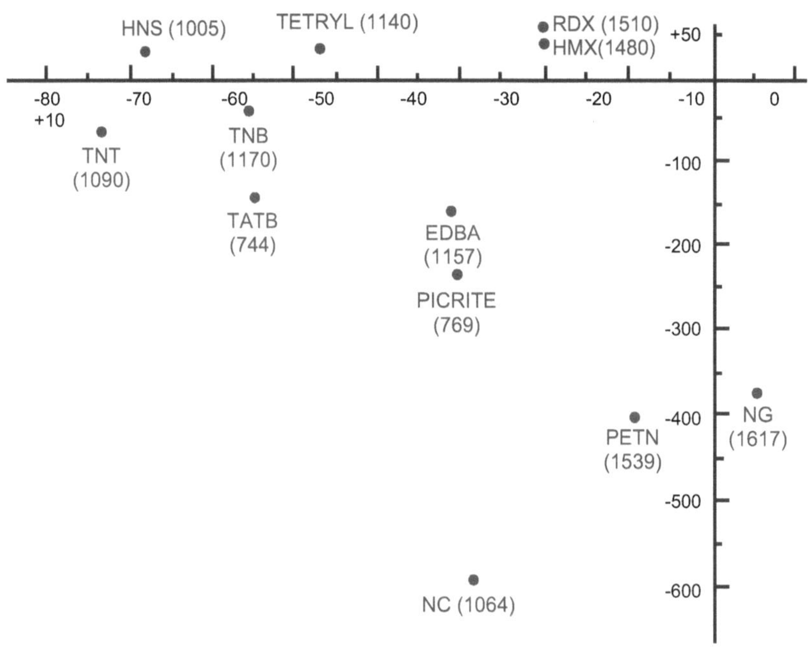

Figure 2.10
Effect of ΔH_f and Oxygen Balance (OB) on ΔH_e (x-Axis: OB%, y-Axis: ΔH_f [cal/g], ΔH_e Values in cal/g Are Given in Brackets).

is important enough to ANFO as to raise the heat of explosion from 0.35 kcal/g for prills alone to 0.89 kcal/g for the oxygen-balanced ANFO. The mixtures used for some common composite explosives and their OB are given in Table 2.2.

2.3.7 Hazard Assessment from OB

In a 1949 *Chemical Reviews* article, W. C. Lothrop and G. R. Handrick demonstrated quantitative correlation between OB and various measures of explosive effectiveness for several classes of organic explosives. This study drew upon the large database accumulated during the years of World War II explosive research. The properties of many explosive compounds were considered and correlated. The authors pointed out that the OB criterion is not only related to the power of new explosive compositions, but it also has a rough bearing on the hazards of their initiation (Table 2.3).

This table shows that as an explosive composition gets closer to zero OB, the hazard of initiation is more.

2.3.8 Composition of Gaseous Products

When an explosive detonates or a propellant burns, it is essential to know the composition of gaseous products formed for calculating ΔH_e and other performance parameters. Because many explosives have negative OB values, during the explosion, there is a stiff competition among C, H, and CO to grab the available oxygen in the explosive molecule.

Table 2.2: Composite explosives and their oxygen balance (OB).

Commercial Name	Composition	Empirical Formula	OB%
AMATOL	80/20 AN/TNT	$C_{0.62}H_{4.44}N_{2.26}O_{3.53}$	1.1193
ANFO	94/6 AN/FO	$C_{0.365}H_{4.713}N_{2.0}O_{3.0}$	−1.6253
COMP A-3	91/9 RDX/WAX	$C_{1.87}H_{3.74}N_{2.46}O_{2.46}$	−50.3723
COMP B-3	64/36 RDX/TNT	$C_{6.851}H_{8.750}N_{7.650}O_{9.3}$	−40.4606
COMP C-4	91/5.3/2,1/1.6 RDX/ di(2-ethylhexyl)sebacate/ polyisobutylene/motor oil	$C_{1.82}H_{3.54}N_{2.46}O_{2.51}$	−46.3755

Table 2.3: Oxygen balance (OB) versus hazard rank.

OB Value	Hazard Rank
More positive than +160	Low
+160 to +80	Medium
+80 to −120	High
−120 to −240	Medium
More negative than −240	Low

Among the possible oxidation reactions (i.e., $H \rightarrow H_2O$, $C \rightarrow CO$, $CO \rightarrow CO_2$), the order of preference appears to depend on the OB of the explosive and, to a certain extent, the density of loading. The situation becomes complex because of the shift in the chemical equilibrium due to side reactions such as the water gas reaction, as follows:

$$CO + H_2O \rightarrow CO_2 + H_2(+9.8\ \text{kcal})$$

$$2CO \rightarrow CO_2 + C(+41.2\ \text{kcal})$$

$$CO + 3H_2 \rightarrow CH_4 + H_2O(+49.2\ \text{kcal})$$

$$2CO + 2H_2 \rightarrow CH_4 + CO_2(+59.1\ \text{kcal})$$

Although databanks and software have recently been developed for the computerized calculation of the exact or at least nearly exact composition of gaseous products, a good approximation by G. B. Kistiakowsky and E. B. Wilson follows to assume the order of preference of these oxidation reactions.

For explosives for which the OB is less than −40%:

$H \rightarrow H_2O$ > $C \rightarrow CO$ > $CO \rightarrow CO_2$

Step 1 Step 2 Step 3

For explosives for which the OB is greater than −40%:

$C \rightarrow CO$ > $H \rightarrow H_2O$ > $CO \rightarrow CO_2$

Step 1 Step 2 Step 3

In the case of detonation of explosives, particularly at higher density, Kamlet and Jacob assume a different order—formation of CO_2 is preferred to the formation of CO. The Kamlet–Jacob method makes this assumption for the estimation of the velocity of detonation (VOD) and the detonation pressure of explosives.

2.3.9 Significance and Limitations of OB

The OB can be used to optimize the composition of the mixture of the explosive. The family of explosives called "amatol" refers to mixtures of AN and TNT. AN has an OB of +20% and TNT has an OB of −74%, which is very deficient in oxygen; therefore, it would appear that the mixture yielding an OB of zero would also result in the best explosive properties. In actual practice, a mixture of 80% AN and 20% TNT by weight yields an OB of +1%, the best properties of all mixtures, and an increase in the strength of 30% over TNT.

The OB provides information on the types of gases liberated. The concept of OB is particularly useful as a first guideline when formulating explosives to produce a minimum of toxic fumes. An explosive with excess oxygen produces toxic nitric oxide and nitrogen dioxide; an explosive with an oxygen deficiency produces toxic CO. Explosives for use underground with poor ventilation should be formulated to produce a minimal total toxic effect. If the OB is large and negative, then there is not enough oxygen for CO_2 to be formed; consequently, toxic gases such as CO will be liberated. This is very important for commercial explosives because the amount of toxic gases liberated must be kept to a minimum.

Sensitivity, brisance (shattering power), and strength are properties resulting from complex explosive chemical reactions; therefore, a simple relationship such as OB cannot be depended upon to yield universally consistent results. When using OB to predict properties of one explosive relative to another, it is to be expected that one with an OB closer to zero will be the more brisant, powerful, and sensitive; however, many exceptions to this rule do exist.

2.3.10 Detonation Temperature/Flame Temperature

The temperature of the gas products on firing propellants in a gun is of considerable importance in the study of ballistics and the erosion of a gun barrel. Likewise, the detonation temperature in the case of high explosives is an important parameter because it is related to the power of those explosives. Let us understand the variation of detonation/flame temperatures under two different conditions—at constant volume and at constant pressure.

Case I (Constant Volume)

When a certain amount of explosive is initiated in a closed vessel that is thermally insulated, let the total heat evolved be x calories. This heat of explosion is used to increase the internal energy of the gases. Because the temperature is effectively a measure of the internal energy of a system, the heat of explosion increases the temperature of the products of explosion. The maximum temperature to which the decomposition products are raised is called the "detonation temperature" in the case of an explosive and the "flame temperature" when we talk about propellants. To be more specific, this temperature is also called the "adiabatic, isochoric flame temperature" (adiabatic, thermally insulated—no heat escapes from or enters inside of the system; isochoric, constant volume), abbreviated as T_V. The isochoric flame temperature of explosives varies from as low as 2500 °C in the case of nitroguanidine to 5000 °C in the case of NG.

Case II (Constant Pressure)

Let us imagine what would happen when the same amount of this explosive is initiated in a vessel that is fitted with a movable piston similar to that in an internal combustion engine. The same amount of heat produced (x calories) heats up the gaseous products to high pressures, but then these gases are now free to move the piston to do some work of expansion. Therefore, only a part of the heat is used to increase the internal energy of the gases (i.e., to the flame temperature), and the rest is converted into work. Obviously, because the amount of heat produced in both cases is the same, the flame temperature in case II—adiabatic, isobaric flame temperature, T_p (isobaric, same pressure)—would be less than T_v.

The above two cases can be written as

$$\text{Case} - \text{I} \quad \Delta H_e = \Delta E_v \quad (\text{Temp}.T_v) \tag{2.3}$$

$$\text{Case} - \text{II} \quad \Delta H_e = \Delta E_p + P\Delta V \quad (\text{Temp}.T_p) \tag{2.4}$$

where ΔE_v and ΔE_p represent the increase in internal energy of the gaseous products at constant volume and constant pressure, respectively. $P\Delta V$ represents the expansion work done by the gases at pressure P to effect an increase in volume by ΔV.

The term $P\Delta V$ is the useful work done by a system, and in the field of HEMs it does the work of blast in the case of high explosives, projectile propulsion in the case of gun propellants, and self-propulsion in the case of rocket propellants.

The relationship between T_p and T_v is as follows:

$$\frac{T_V}{T_P} = \gamma \tag{2.5}$$

where γ is the mean molar value of the ratio of specific heats of product gases at constant pressure (C_p) and at constant volume (C_v) (i.e., C_p/C_v of the product gases).

2.3.10.1 Calculation of Detonation/Flame Temperature

Let us assume that during an explosive reaction, n_1, n_2, and n_3 moles of CO, $H_2O_{(v)}$, and CO_2 are produced, respectively, and the flame temperature is T_v. After the heat of explosion (ΔH_e) is released, the gases cool to ambient temperature (T_a). This can be represented as

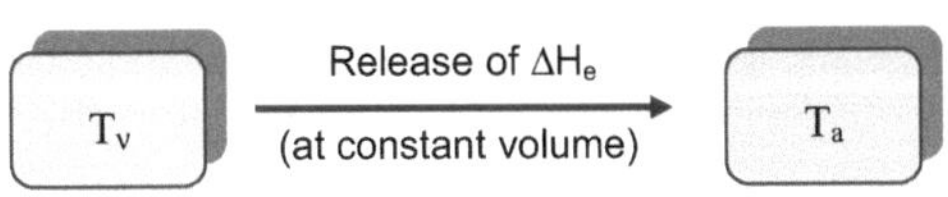

Conversely, we can imagine that the above gases are heated from T_a to T_v using the heat, ΔH_e. The amount of heat needed to heat each gas is obtained by multiplying the number

of moles of the gas produced, its molar heat capacity, and the increase in temperature. If $(C_v)_{CO}$, $(C_v)_{H_2O}$, and $(C_v)_{CO_2}$ are the molar heat capacities of CO, H_2O, and CO_2, respectively, then it can be written

$$\Delta H_e = n_1 (C_v)_{co}[T_v - T_a] + n_2 (C_v)_{H_2O}[T_v - T_a] + n_3 (C_v)_{CO_2}[T_v - T_a]$$

That is, $\Delta H_e = \Sigma C_v \times (T_v - T_a)$, where ΣC_v is the mean molar heat capacity of the product gases.

This above equation can be rearranged as follows:

$$T_v = \frac{\Delta H_e}{\sum C_v} + T_a \qquad (2.6)$$

Because T_a and ΣC_v are constants, it is seen from Eqn (2.6) that T_v linearly increases with ΔH_e. This is illustrated in the worked example given below.

2.3.10.1.1 Calculation of T_v from Molar Internal Energies of the Products of Explosion

Standard tables are available (refer to *Explosives*, by Rudolf Meyer, 4th ed., Table No. 35) that give the molar internal energies of the reaction products in relation to temperature (Table 2.4). The best way to calculate T_v is to plot the calculated heat of explosion against various temperatures using the above table. From the linear plot, we can find out the value of T_v knowing the experimental value of ΔH_e.

Worked Example 2.2

Calculate the isochoric and isobaric flame temperatures of PETN.

(Given: The heat of explosion of PETN = 1510 cal/g)

Table 2.4: Molar internal energies of products $C_v(T - T_a)$; T_a = 25 °C (~ 300 K).

	Molar Internal Energies of Explosion Products (kcal/mol)			
Temperature (K)	**N_2**	**H_2O**	**CO**	**CO_2**
2500	13.15	18.43	13.33	24.34
3000	16.57	23.81	16.78	30.81
3500	20.05	29.37	20.27	37.43
4000	23.79	35.03	23.79	44.13
4500	27.08	40.76	27.33	50.88
5000	30.62	46.54	30.88	57.67

Reproduced with permission from: R. Meyer, J. Kohler, Explosives, VCH Publishers, Germany, 1993.

PETN, $C(CH_2ONO_2)_4$, or $C_5H_8N_4O_{12}$, undergoes the following explosive reaction:

$$\underset{\text{(Mol.Wt=316.1)}}{C_5H_8N_4O_{12}} \rightarrow \underset{\text{(Total: 11 moles of gases)}}{2N_2 + 4H_2O_{(v)} + 2CO + 3CO_2}$$

We need the heat of explosion value in the unit of kilocalories per mol.

$$\Delta H_e = 1510\ \text{cal/g} = \frac{1510}{1000} \times 316.1 = 477.3\ \text{kcal/mol}$$

The minimum and maximum values of flame temperature of explosives are approximately 2500 and 5000 K, respectively. We do not know the actual flame temperature of PETN, although we are certain that it should be somewhere between 2500 and 5000 K. Using Table 2.4, we can calculate the expected ΔH_e values of PETN had its flame temperature been 2500, 3000, 3500, 4000, 4500, or 5000 K.

For example, at 2500 K (or had the flame temperature been 2500 K), the expected cal.val output by the products $2N_2 + 4H_2O + 2CO + 3CO_2$ would be

$$\begin{aligned} \Delta H_{e(2500)} &= 2(13.15) + 4(18.43) + 2(13.33) + 3(24.34)\ \text{kcal/mol} \\ &= 199.70\ \text{kcal/mol.} \end{aligned}$$

A similar calculation yields ΔH_e values of 254.37, 310.41, 367.67, 424.50, and 482.17 kcal/mol at 3000, 3500, 4000, 4500, and 5000 K, respectively. A plot of cal.val versus assumed T_v (see Figure 2.11) yields a straight line.

Because the experimentally determined value for ΔH_e is 477.3 kcal/mol, it can be read out from the plot that the actual value for T_v is approximately 4960 K.

2.3.10.1.2 Calculation of T_p

Because T_p and T_v are related as $\frac{T_v}{T_p} = \gamma$, we should calculate the molar average value of γ for all of the products. The values of γ for N_2, H_2O, CO, and CO_2 are 1.404, 1.324, 1.404, and 1.304, respectively. The molar average of the products can be written as (remember that there are 11 moles of the product gases in all)

$$\begin{aligned} \gamma &= \left(\frac{2}{11} \times 1.404\right) + \left(\frac{4}{11} \times 1.324\right) + \left(\frac{2}{11} \times 1.404\right) + \left(\frac{3}{11} \times 1.304\right) \\ &= 1.348 \end{aligned}$$

$$T_p = T_v/\gamma = 4960/1.348 = 3680\ K$$

The above method of calculating flame temperature can be applied to compositions of explosives and propellants once we know their ΔH_e values and the composition of the gaseous products.

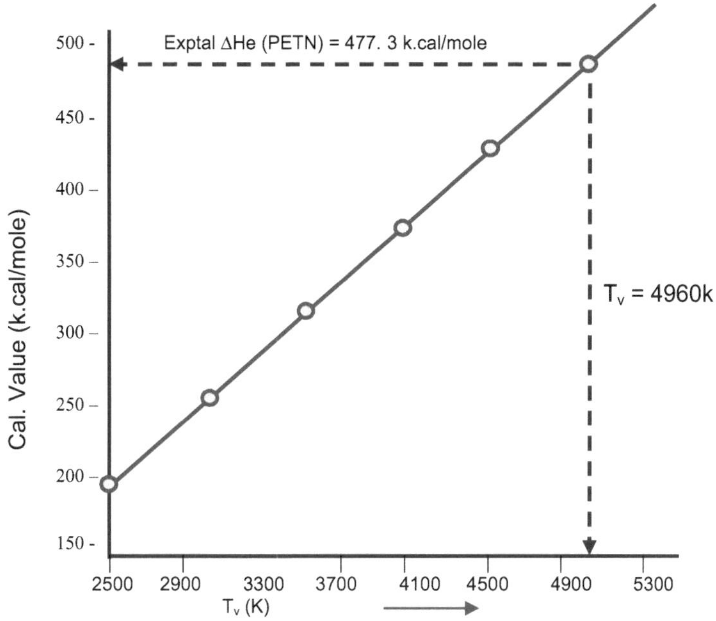

Figure 2.11
Plot of Calculated cal.val versus Different T_V Values.

2.3.10.1.3 Effects of C_V Values

There is an interesting observation that during an explosive decomposition, if the product gases have smaller molecular weights, then the flame temperature marginally increases. The smaller the molecule, the lesser is its heat capacity and, as a result (because $T_V = \Delta H_e/\Sigma C_V + T_a$), the flame temperature marginally increases.

2.3.10.1.4 Value of γ

The value of γ (of the product gases), which is the ratio of C_p to C_V, plays an important role in determining the energetic parameters of explosives and propellants. γ decreases with increasing temperature but increases with pressure. However, in a process of explosion/propellant burning, which is a high-temperature/high pressure phenomenon, this increase/decrease is almost compensated for and, with reasonable approximation, one can use the γ values of the product gases given for room temperature and ambient pressure at the conditions of explosion. This value is approximately 1.3−1.4 for most of the CHNO explosives.

However, at detonation/shock-wave zones in which the pressure ranges are phenomenally high, on the order of several hundreds of thousands of atmospheres, the value of γ sharply increases to approximately 3.

Note: "Detonation temperature" and "flame temperature" are almost the same for a given HEM because they refer to the temperature to which the products of explosion are adiabatically heated by the heat of explosion. However, the term "explosion temperature" (also sometimes referred to as "cook-off temperature") is often used to refer to the temperature at which the autoignition of an explosive commences when it is heated at a particular rate. For example, the detonation temperature of NC is approximately 3470 K whereas its explosion temperature is approximately 170 °C when it is heated at the rate of 5 °C/s. That is, when the temperature of NC reaches approximately 170 °C, the autoignition starts. There can be a slight variation in the values of the explosion temperature of an explosive depending on the heat exchange conditions and the geometry of the sample.

2.3.11 Gas Volume

When a certain quantity of explosive undergoes an explosive decomposition, it evolves high-pressure/high-temperature gaseous products. Because of the high pressure, the gases expand to reach the atmospheric pressure, and in the process of expansion, they do work. Because the volume of a (solid) explosive is negligible in comparison to that of the product gases, we can write

$$PV = nRT \tag{2.7}$$

where P, V, n, R, and T represent the final pressure after expansion, the final volume, the number of moles of the gases produced, the universal gas constant, and the final temperature, respectively. The volume of the gaseous products of expansion (V) is generally calculated at the pressure of 1 bar and 273 K (i.e., at normal temperature and pressure (NTP)). For explosives, the value of V varies from 700 to 1000 cc/g. That means that in the case of most of explosives, the explosion of 1 g of an explosive produces product gases that occupy a volume varying between 700 and 1000 cc when measured at atmospheric pressure and 273 K.

Worked Example 2.3

Calculate the number of moles and the volume of the gaseous products of explosion of RDX ($C_3H_6O_6N_6$).

The explosive reaction of RDX is given as

$$C_3H_6O_6N_6 \rightarrow 3CO + 3H_2O_{(v)} + 3N_2$$

$$(\text{Mol.Wt} = 222)$$

There are 9 moles of gaseous products, including H_2O, which is in vapor state. As a standard practice, H_2O is treated as vapor even when we calculate the total gas volume at NTP.

$$222\text{ g of RDX evolves} \rightarrow 9\text{ mol of gases (at NTP)}$$

$$\text{Therefore, 1 g of RDX evolves} \rightarrow 9/222\text{ mol of gases at NTP.}$$

$$\text{(Applying Avogadro's law)} \rightarrow \frac{9}{222} \times 22,400\text{ cc of gases at NTP} = 908\text{ cc.}$$

The gas volume of RDX explosion products = 908 cc/g.

It is seen from the gas equation that at a given temperature and pressure, the volume of a given gas directly depends on the number of moles of the gaseous products. Because the volume generation is tantamount to the work of expansion, we can say that an explosive that on decomposition produces more moles of the product gases (per gram of the explosive) possesses better work potential. More moles of gases per gram of explosive effectively means the product gases with lesser molecular weights.

2.3.12 The nRT Wonder

In Section 2.3.10, we presented the equation relating ΔH_e and ΔE at constant pressure as

$$\Delta H_e = \Delta E + P\Delta V$$

ΔV refers to the change in volume when a solid explosive is converted into product gases. As compared to the volume of product gases, the volume of the solid explosive can be neglected (we have seen above that 1 g of RDX, which occupies a volume of 0.56 cc, on explosion, gives product gases that occupy a volume of 908 cc). Therefore, in the above equation, ΔV can be replaced by V, the volume of the product gases; that is, $\Delta H_e = \Delta E + PV$, and because $PV = nRT$ (assuming ideal gas behavior),

$$\Delta H_e = \Delta E + nRT \tag{2.8}$$

nRT is actually the work factor of an explosive decomposition. This term is very important in the field of HEMs, and it manifests its importance in different forms under different nomenclatures. As we will see in the respective chapters, the nRT factor manifests itself as

- Specific energy, which decides the strength or power of a high explosive;
- Impetus, or force constant, in gun propellants, which would decide how much muzzle velocity and hence range can be imparted to a projectile; and
- A parameter in rocket propellants that is directly related to the specific impulse (I_{sp}), the ultimate energy index for any rocket propellant.

Although in the first two cases we deal with an almost constant-volume condition (flame temperature: T_v), in the case of rocket propellants we encounter a constant-pressure condition (flame temperature: T_p). These will be discussed in a little more detail in the respective chapters. The message is: "If we want an HEM with better work potential, then the value of nRT must be higher, implying that for a given weight of the explosive/propellant, it should produce more moles of product gases with higher flame temperature (isochoric or isobaric depending on the function)."

Energy of Formation (ΔE_f) versus Heat of Formation (ΔH_f):

We defined ΔH_f and explained its importance in Section 2.3.2. Now, having understood the difference between ΔH and ΔE being the energy transition involved under constant pressure and constant volume, respectively, let us see the relation between the energy of formation (ΔE_f) and the enthalpy of formation (ΔH_f).

"ΔH_f and ΔE_f are the quantities of heat absorbed or evolved when 1 mole of a compound is formed from its constituent elements at standard state (25 °C and 1 atm) at constant pressure and constant volume, respectively."

Worked Example 2.4

The enthalpy of formation of RDX is 76.1 cal/g. Calculate its energy of formation.

(Given: RDX: $C_3H_6N_6O_6$; molecular weight = 222.1.)

The chemical equation for the formation of RDX can be written as

$$3C + 3H_2 + 3N_2 + 3O_2 \rightarrow C_3H_6O_6N_6$$

Because C (carbon) and $C_3H_6O_6N_6$ (RDX) are solids under standard states, the change in the number of moles of gaseous compounds will be

Δn = moles of the gaseous products − moles of the gaseous reactants $= 0 - (0 + 3 + 3 + 3) = -9$.

Because $\Delta H = \Delta E + \Delta nRT$ (R = universal gas constant = 1.987 cal/K/mol), and T = standard temperature = 25 °C = 298 K, then we can write $(76.1 \times 222.1) = \Delta E + (-9)(1.987)(298)$. (Please note that cal/g must be converted into cal/mol by multiplying by the molecular weight of RDX).

$16{,}902 = \Delta E - 5329$.

Therefore, $\Delta E = 22{,}231$ cal/mol.

$$= \frac{22{,}231}{222.1} \approx 100 \text{ cal/g}$$

Energy of formation (ΔE_f) of RDX = 100 cal/g.

On the basis of the data on the ΔE_f values of explosives and their explosive decomposition products, we can calculate their heat of explosion and perform a thermodynamic calculation of the decomposition reactions.

2.3.13 Pressure of Explosion

It was stated that when an explosive undergoes deflagration in a closed vessel, high pressure is produced because of the evolution of high-temperature gases in large amounts. This pressure is an important parameter because when the product gases expand to do some useful work, such as propulsion of a projectile through a gun barrel, the total amount of work done by the gases is directly proportional to this pressure. The pressure of explosion, P_e, is defined as the maximum static pressure achieved when a given weight of explosive is burned in a closed vessel of fixed volume assuming adiabatic conditions. The gas equation for this process is given as

$$P_e(V^* - \alpha) = nRT_e$$

where V^* is the volume of the closed vessel and α is the covolume correction necessitated by the fact that at such high pressures, a gas tends to be nonideal and a certain correction must be applied for the volume of gaseous molecules themselves. We will deal in more detail in subsequent chapters about the nonideal behavior of gaseous products formed during explosion during detonation and explosive deflagration.

(Note: The pressure of explosion should not be confused with detonation pressure. The latter refers to the pressure that exists at the *detonation zone* (detonation front) when a shock wave travels through the medium of the explosive, which will be discussed in the next chapter.)

2.3.14 Density

Density is one of the important characteristics of explosives and propellants. It will be shown later that an increase in density of a high explosive increases its VOD and brisance (destructive fragmentation effect). The actual density of an explosive, referred to as the "theoretical maximum density" (TMD), can be accurately determined by conventional methods. However, when an explosive composition is processed and filled, say in a warhead, the density of the composition is often slightly less because of very fine voids. That is why maximum care is taken to maximize the density of the high-energy composition to be close to the TMD.

In addition, in the case of propellants, the higher the density, the higher will be the performance output. For example, if the volume of a cartridge case of a small arms ammunition is limited, one would look for a propellant with high density so that more weight of the propellant can be loaded in it. In the case of rocket propellants, even if a solid rocket propellant may be energetic, if its density is very low, then the weight of the rocket propellant grain loadable in a rocket motor of limited volume will be too little to be acceptable. Therefore, in the field of energetic materials, density is a parameter as important as energy itself.

2.3.14.1 Density and Molecular Structure

The density of an explosive should depend on the nature of the molecules and the way they are arranged or packed in a crystal lattice. In particular, the weight of a molecule and its volume (effective molar volume) should be a dominant factor. L. T. Eremenko established a linear relationship between the density of explosives (liquid and solid) and their hydrogen content, classified the explosives under 12 groups depending on their molecular structures (whether aliphatic or aromatic, with symmetrical or unsymmetrical substituents, etc.), and evolved an empirical equation as follows:

$$\rho = a_i - K_i H$$

where ρ represents the calculated density of the explosive at TMD. a_i and K_i are constants, the values of which depend on the molecular structure/group/homologue. H refers to the weight percentage of hydrogen in the molecule (normally %H is from 0 to 6). An error of not more than 2% in this method has been claimed.

To summarize, several vital parameters of HEMs dictate their ultimate performance characteristics. The inter-relationships among them are schematically shown in Figure 2.12.

Summary of Important Terms

1. Heat of reaction
   The quantity of heat evolved or absorbed during a chemical reaction is called the "heat of reaction."
2. Enthalpy of reaction
   If the chemical reactions occur at constant pressure, then the heat of reaction is often called the "enthalpy of reaction."
3. Endothermic reaction
   A reaction in which energy is supplied to the reactants from the surroundings to obtain the product is called an "endothermic reaction."
4. Exothermic reaction
   A reaction in which heat energy is evolved along with the products is called an "exothermic reaction."
5. Heat
   Heat is one form of energy and can be produced from work. However, it is not completely convertible into work. It can only partly be transformed into work. In this respect, heat differs from many other forms of energy.

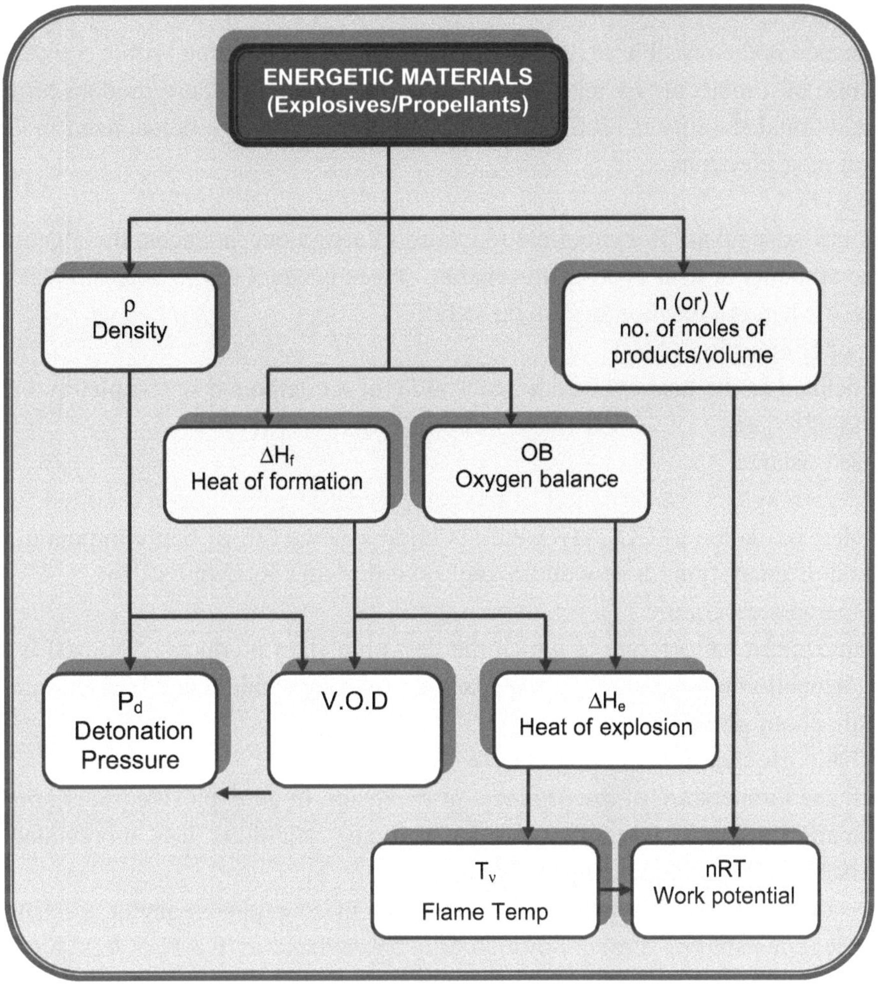

Figure 2.12
Inter-Relationship between Parameters and Performance Characteristics of High-Energy Materials.

6. Energy
   The energy of a system may be defined as "any property that is capable of doing work." There are several forms of energy, including thermal energy (heat), mechanical energy, electrical energy, chemical energy, etc. Energy can be quantitatively converted into work and can be produced from work.
7. Internal energy
   Internal energy is the total energy content of the system. It is due to the translational, vibrational, and rotational motions of the molecules and their mutual attraction (intermolecular force) in a system.

8. Resonance
   Resonance is the possible existence of several types of bonding within a fixed skeleton structure of a molecule by the mobility of double bonds. In more modern terminology, the additional stability is brought about by the formation of a delocalized molecular orbital of π electrons.
9. Hess's law
   Hess law states that "If a chemical reaction is carried out in stages, the algebraic sum of the amounts of heat evolved in separate stages is equal to the total amount of heat evolved when the reaction occurs directly."
10. Heat of combustion (ΔH_c)
    It is defined as the heat evolved when 1 mole of a compound is completely burnt in excess of oxygen.
11. Oxygen balance
    The percentage excess or deficit of oxygen present in a compound required for its complete oxidation to CO_2, H_2O, etc., is known as the OB of that compound. OB is a method of quantifying how well an explosive provides its own oxidant.
12. Detonation temperature
    The maximum temperature to which the decomposition products are raised is called the "detonation temperature" in the case of explosives and "flame temperature" when we talk about propellants.
13. Adiabatic, isochoric flame temperature
    The flame temperature of the products of explosion of an explosive under adiabatic (thermally insulated), isochoric (constant volume) conditions. It is abbreviated as T_V.
14. Adiabatic, isobaric flame temperature
    Flame temperature of the products of explosion of an explosive under adiabatic, isobaric (constant pressure) conditions. It is abbreviated as T_P. T_P is less than T_V.
15. Explosion temperature/autoignition temperature
    The temperature at which the autoignition of an explosive commences when it is heated at a particular rate.
16. Pressure of explosion P_e
    The pressure of explosion (P_e) is defined as the maximum static pressure achieved when a given weight of explosive is burned in a closed vessel of fixed volume assuming adiabatic conditions.

Suggested Reading

Any standard book on Physical Chemistry would discuss various aspects of thermochemistry. Apart from this, the reader might refer to the following books.

[1] A. Bailey, S.G. Murray, Explosives, Propellants, and Pyrotechnics, Pergamon Press, Oxford, New York, 1988.
[2] Service Textbook of Explosives, Min. of Defence, Publication, UK, 1972.

[3] Structure and properties of energetic materials, in: D.H. Liebenberg, et al. (Eds.), Materials Research Society, 1993. Pennsylvania, USA.
[4] P.W. Cooper, Explosives Engineering, VCH, Publishers Inc., USA, 1996.
[5] B. Siegel, L. Schieler, Energetics of Propellant Chemistry, John Wiley & Sons. Inc., New York, 1964.
[6] S.F. Sarner, Propellant Chemistry, Reinhold publishing corporation, New York, 1966.
[7] L. Pauling, Nature of the Chemical Bond, third ed., Cornell University Press, Ithaca, 1960.

Questions

1. The heat of explosion of TNT is 1080 cal/g. If 1 kg of TNT detonates in 2 μs, how much power does it generate? (Answer: 2.2572×10^{12} W)
   (Note: The above question is hypothetical. The Second Law of Thermodynamics is very much there to ban us from converting the entire heat to useful work.)
2. Why can we describe explosives as metastable materials?
3. When an explosives chemist wants to synthesize a new, high-performing explosive, what parameters should his target molecule satisfy?
4. Why does one prefer to have a HEM with a positive heat of formation?
5. Calculate the OB of PETN. (Answer: 60.76%)
6. An explosive has a unique value of heat of combustion whereas its exact value of heat of explosion depends on the conditions of its experimental determination. Why?
7. Why does zero OB help to achieve highest value of heat of explosion?
8. What is meant by isochoric and isobaric flame temperatures (T_v and T_p respectively)? How are they related to each other? Why is T_v always more than T_p?
9. Calculate the isochoric and isobaric flame temperatures of HMX (molecular formula $C_4H_8N_8O_8$). (Given: Heat of explosion of HMX = 1480 cal/g.) (Hint: Use the molar internal energies table given in Section 2.3.7.1.)
   (Answer: $T_v \sim 4580$ K, $T_p \sim 3326$ K.)
10. What is the importance of the gas volume for an explosive? Calculate the volume of the gaseous products of the explosion of 1 g of NG (molecular formula $C_3H_5N_3O_9$) measured at NTP. Assume water as water vapor. (Answer: 715.1 mL)
11. Which parameter decides the work potential of an HEM? What different names does it assume for a high explosive, a gun propellant, and a rocket propellant?
12. Name some methods used for the theoretical prediction of the heat of formation of a molecule.
13. What is the general method used to measure the detonation temperature of a high explosive?
14. What is the difference between explosion temperature and flame temperature?
15. Define heat of formation (ΔH_f) and energy of formation (ΔE_f) and state how they are related to each other.
16. Define pressure of explosion (P_e) and write the gas equation for this process.

CHAPTER 3

Two Faces of Explosion: Deflagration and Detonation

3.1 Explosion

Explosion is one of the most common words used in our day-to-day life. "Bursting" and "detonation" are the words that appear to be synonymous with the word "explosion." In a way, "explosion" is a loosely used word implying different meanings under different situations. We say, "a balloon explodes," "a warhead or a bomb explodes," "a nuclear weapon explodes," "a gas cylinder explodes," "a reaction vessel explodes," and so on (not to mention its figurative usage such as "a boss explodes" or "a wife explodes"). although all of these situations vary greatly in terms of

- The type of energy release (physical/chemical/nuclear)
- The quantum of energy released, and
- The rate at which the energy is released.

Before we try to get the correct definition of the word "explosion," let us see how an explosion can be classified. There are three types of explosions: (1) physical explosions, (2) chemical explosions, and (3) nuclear explosions.

Physical explosions involve very fast physical transformation of a system or material that results in an explosion. An example is the explosion of an overheated water boiler. No chemical change takes place in this process. Only water in the liquid state gets converted into its vapor state. Because the water vapor occupies a much larger volume than liquid (water) at its boiling point, the pressure developed by the water vapor in a confined volume of the boiler is so high that it overcomes the strength of the container material, leading to sudden energy release.

On the other hand, nuclear explosions are disastrous because of the enormous amount of thermal energy and radioactivity released due to the conversion of mass into energy obeying the well-known equation of Einstein, $E = mc^2$.

This chapter excludes the above two types of explosions (physical and nuclear) and deals only with chemical explosions, in which a large amount of thermal energy, often accompanied by the evolution of a large amount of high-pressure, high-temperature gaseous products, is suddenly released because of a chemical reaction. Coming back to the

Demystifying Explosives: Concepts in High Energy Materials. http://dx.doi.org/10.1016/B978-0-12-801576-6.00003-3

definition of the term "explosion," unfortunately, no definition is perfect in revealing all of the characteristics of an explosion. The most acceptable one can probably be stated as follows: "Explosion is a process of rapid physical or chemical transformation of a substance, accompanied by *an extremely rapid transition of its potential energy into mechanical work.*" A chemical explosion can be subclassified as follows:

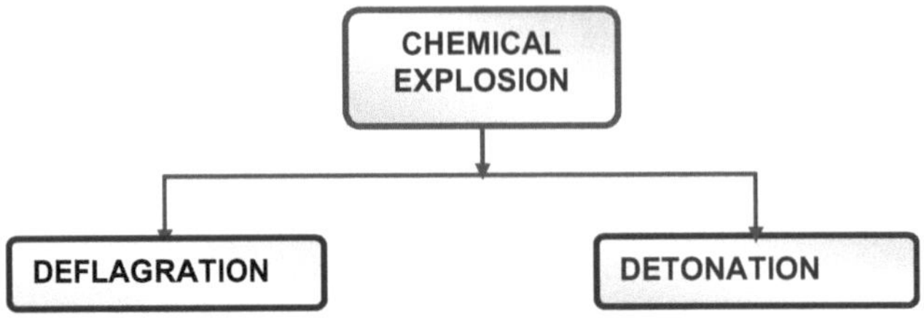

Most of the chemical explosions involve rapid chemical reactions, as a result of which large volumes of high-pressure and high-temperature gases are formed in a short time with the evolution of an enormous quantity of heat. For example, explosion of RDX (cyclotrimethylene trinitramine) is accompanied by the evolution of 9 mole of gaseous products in a few microseconds.

$$C_3H_6N_6O_6 \rightarrow 3CO_{(g)} + 3H_2O_{(g)} + 3N_{2(g)} + \text{Heat}$$

In rare cases, no or very little gaseous products are evolved during a chemical explosion. For example, the explosion of copper acetylide is as follows:

$$Cu_2C_2 \rightarrow 2Cu + 2C + \text{Heat}$$

Here, the reactant and the products are solids. There are no gases.

In addition, when a mixture of hydrogen and oxygen explodes to produce water, there is actually a reduction in volume:

$$2H_{2(g)} + O_{2(g)} \rightarrow 2H_2O_{(g)} + \text{Heat}$$

These can be explained by the fact that both of these reactions are highly exothermic and a large amount of heat is released in a very short time, thereby suddenly heating up the adjacent gases or air and creating high-pressure waves or shock waves.

3.2 Deflagration and Detonation

Explosives are those substances that have their own supply of oxygen in their molecules. When they are initiated, they may either burn violently (deflagrate) or explode disastrously generating shock waves (detonate). What are the differences between deflagration and detonation?

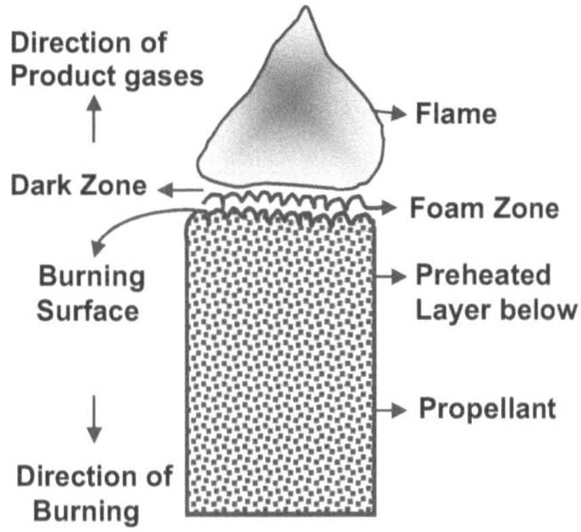

Figure 3.1
Deflagration of a Propellant Stick.

Let us take a stick of a rocket propellant, say, made of nitrocellulose (NC) and nitroglycerine (NG; i.e., a "double-base" propellant). When it is ignited at one of its ends, it burns rather vigorously, layer by layer. The salient points of a deflagration process are indicated in Figure 3.1.

Deflagration has the following characteristics:

1. The propellant burns layer by layer.
2. There are different zones existing above the burning surface as shown, varying in temperature, pressure, concentration, and composition of gaseous products.
3. The hot gaseous products emerge away from the regressing surface.
4. The most important characteristic of deflagration is that the rate of deflagration (or the rate of recession of a burning surface, often expressed in millimeters per second at a given pressure) is much below the sonic velocity of the material (i.e., the velocity of sound through the propellant material).
5. The process of deflagration is sustained by thermal feedback from the flame to the surface temperature by means of conduction, convection, and radiation.
6. The rate of regression (or burning rate, r) heavily depends on the pressure of the surrounding gases (P), and, according to Vielle's law, a double-base propellant nearly obeys the equation

$$r = \mathrm{b}\mathrm{P}^{\mathrm{n}}$$

   where n is the pressure exponent and b is a constant. The value of n depends on the propellant composition, the pressure, etc., the details of which will be seen in subsequent chapters on propellants.

Let us see what happens when the process of detonation occurs in an explosive.

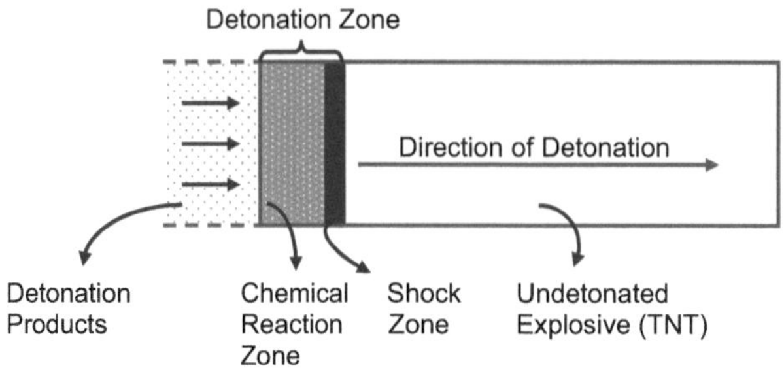

Figure 3.2
Detonation of an Explosive.

When a cylindrical stick of trinitrotoluene (TNT) is detonated using a detonator, the following characteristics are noted during the process of detonation (see Figure 3.2):

1. The detonation is accompanied by the production of a shock wave.
2. The wave front of the shock wave has a high temperature and pressure gradient (shock zone), which instantaneously initiates chemical decomposition of the shocked explosive layer of the undetonated explosive. The chemical reaction of explosion is completed in the chemical reaction zone. The shock zone is very narrow ($\sim 10^{-5}$ cm) as compared with the chemical reaction zone (varies from 0.1 to 1.0 cm), and both of these zones together form the detonation zone.
3. The gaseous products flow in the same direction as that of the propagation of detonation.
4. The rate of propagation of the detonation front (velocity of detonation (VOD)) is more than the sonic velocity of the material (i.e., the velocity of sound in undetonated TNT). The VOD varies from 1500 to more than 9000 m/s for different explosives.

The important differences between deflagration and detonation are summed up in Table 3.1. The actual nature of a shock wave will be discussed in the subsequent section of this chapter.

3.3 Linear Burning and Mass Burning

In an earlier chapter, it was mentioned that when a chunk of coal burns in air, it takes its own time; however, when it is powdered to very fine dust, dispersed in air, and ignited, a violent detonation (that takes <1 ms) results. Has the basic characteristic of burning of this piece of coal changed after being reduced to fine dust? No. The chemistry of combustion of coal in air remains the same. If a given sample of coal burns at the rate of, say 1 mm/s under atmospheric pressure (what we call the "linear burning rate"), then this property does not change irrespective of how small you break it into pieces or grind to a fine dust. The finer it is, then the more surface area is exposed for burning. Let us assume

Table 3.1: Comparison of deflagration vs detonation.

Sl. No.	Deflagration	Detonation
1.	It is a surface phenomenon (i.e., its propagation is by layer-to-layer burning).	It is a shock-wave phenomenon (i.e., high-speed shock wave traveling through the explosive medium propagates detonation).
2.	The rate of deflagration is lower than the sonic velocity in the medium.	The rate of detonation is higher than the sonic velocity in the medium.
3.	The products of deflagration go away from (opposite to) the direction of propagation of deflagration.	The products of detonation travel in the same direction as that of the propagation of detonation.

that, on average, after fine division, each coal particle (assuming to be a sphere) has 1 μm (10^{-3} mm) as its radius. At the linear burning rate of 1 mm/s, each particle will take only 10^{-3} s (i.e., 1 ms) for burning. Because there are so many millions of such particles, a great amount of pressure is developed in 1 ms, when all of these particles undergo simultaneous ignition/burning. This results in a sudden increase in pressure (even before each particle is fully consumed), and the huge pressure converts the sound wave into a shock wave, resulting in detonation. Here, we talk about the "mass burning rate," which tells us how many grams of the material will be consumed per unit time. It depends on the linear burning rate of the material (r), the surface area exposed for burning (A), and the density of the material (ρ).

The mass burning rate, $\dot{m}$ is related to these parameters as follows:

$$\dot{m} = rA\rho$$

Please note that r and $\dot{m}$ have units of millimeters per second (or cm/s) and grams per second (or kg/s), respectively. The purpose of introducing the concept of linear burning and mass burning at this stage is because it plays a key role in propellant ballistics and in phenomena such as deflagration-to-detonation transition (DDT), which will be separately discussed.

3.4 Shock Wave and Detonation Wave

A shock wave is a disturbance propagating at supersonic speed in a material, accompanied by an extremely rapid increase in pressure, density, and temperature. When a large amount of energy is suddenly released in a very limited space, it produces a shock wave. It may be mechanical energy (e.g., passage of a supersonic aircraft), electrical energy (e.g., discharge of lightning in a narrow channel), or chemical energy (e.g., detonation of an explosive). The shock wave caused by a detonation is called a "detonation wave." Therefore, a detonation wave is a shock wave, but all shock waves are not detonation waves. When a shock wave is not sustained, say, by continuous feeding of energy (as in case of a detonation wave, in which continuous evolution of thermochemical energy and gaseous products behind the shock front keeps feeding the shock wave), it loses energy because of viscous dissipation by the surrounding medium and it degenerates into a sound wave (e.g., thunder).

The detonation process in an explosive requires a shock wave for initiation. This shock wave that initiates a detonation may originate from the detonation of an explosive nearby (sympathetic detonation) or from a process of deflagration (which is subsonic) that gets transformed into a supersonic disturbance because of reasons such as confinement. The shock wave in all of these cases should be supersonic. It compresses, heats, and ignites an explosive that gives out sufficient energy and expanding reaction products to sustain the shock wave.

3.4.1 The Concept of a Shock Wave

The formation of a one-dimensional planar shock wave can be visualized with the help of an accelerating piston in small increments from zero velocity to some final constant velocity (see Figure 3.3(a)–(d)).

Figure 3.3(a) shows that the first infinitesimal compression at the piston face results in the propagation of a sound wave (velocity = C_o). In Figure 3.3(b), the material that is in a compressed state has a higher density and the velocity of sound in this denser medium is more than C_o (e.g., = C_1). This means that the wave front C_1 will catch up with the wave front C_o after a particular time. Because the acceleration of the piston is continuous, it can be imagined that the medium facing the piston gets more and more compressed, resulting in a train of waves in which the first is at the speed of sound in the undisturbed material (C_o), followed by faster and faster moving wave fronts of higher and higher pressures.

From Figure 3.3(c), it can be visualized that after some time C_1 catches up with C_o, then C_2 catches up with C_1, and so on, so that eventually all of the waves coalesce into a single, steep, discontinuous wave front across which exists a sharp discontinuity in pressure, density, and temperature (Figure 3.3(d)). The width of this discontinuity is generally on the order of a few molecular mean-free-path lengths. Behind the piston, a reverse process of gas expansion creates a rarefaction wave that moves in a direction opposite to the shock wave and piston motion. There is a drastic change in the physical properties of the medium across the shock front (Figure 3.4). This change is described by the Rankine–Hugoniot (RH) equations, ensuring the following:

1. Conservation of mass;

$$\frac{V_o - V}{V_o} = \frac{U_p - U_o}{U_s} \tag{3.1}$$

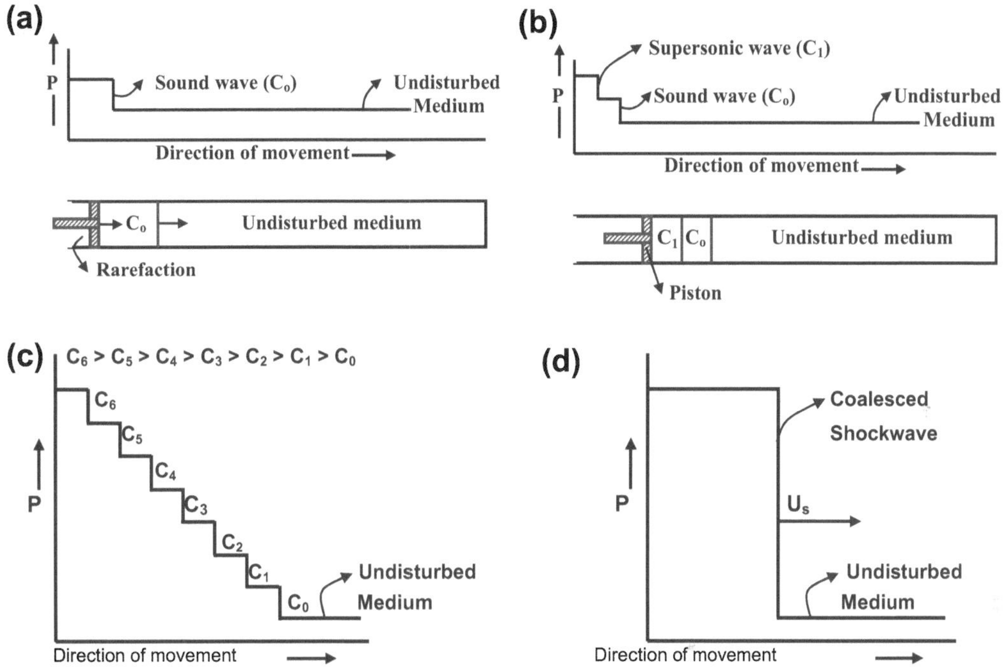

Figure 3.3

Formation of a Shock Wave at Time (a) t_o, (b) t_1, and (c) t_6. (d) Coalescence of Wavefronts to a Plane Shock Wave. (*Note*: The Velocity of Sound in a Medium is Given as $C = (\gamma RT_0)^{1/2}$, where γ is the Ratio of the Specific Heat of the Medium, T_o is its Absolute Temperature, and R is the Universal Gas Constant. During Compression, the Medium Gets Heated up and the Value of T_o Increases, Thereby Increasing the Velocity of Sound in the Medium.)

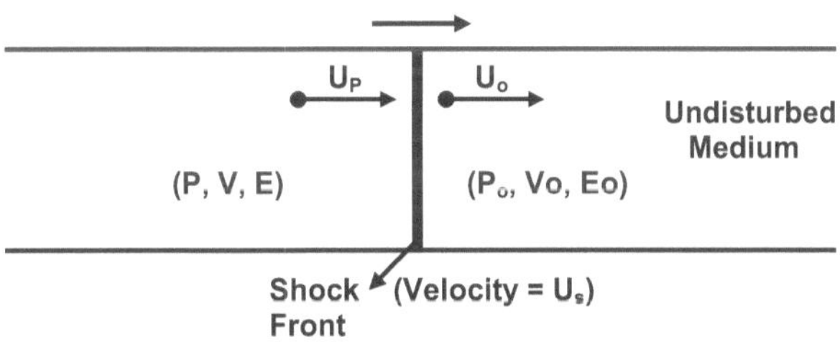

Figure 3.4

Movement of Shock Front.

2. Conservation of momentum; and

$$P - P_o = \frac{U_s(U_p - U_o)}{V_o} \tag{3.2}$$

3. Conservation of energy

$$E - E_o = \frac{(P + P_o)(V_o - V)}{2} \tag{3.3}$$

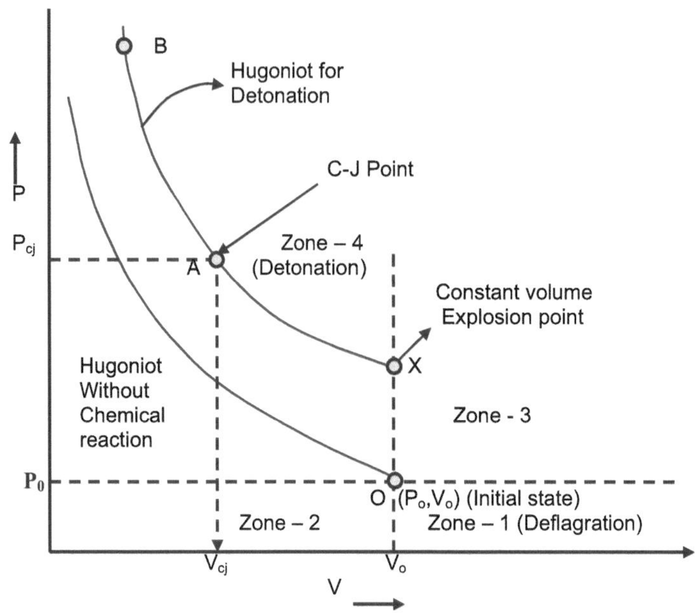

Figure 3.5
Hugoniot Curve for Detonation.

where U_S is the shock velocity and E, V, P, and U_P are, respectively, the energy, specific volume (i.e., volume occupied by 1 g of the substance), pressure, and material (or piston) velocity in the shocked states. The subscript zero indicates the initial state. The RH curve represents the locus of all final states that can be reached by shock-compressing a material from the same initial state. The resultant curve of pressure against volume is known as the Hugoniot curve (Figure 3.5). If the initial state is known, by measuring any two of the five final parameters, then the final state properties can be determined. The shock velocity (U_S) is usually measured.

The lower P—V curve in Figure 3.5 represents a simple Hugoniot curve for inert material that does not involve any chemical reaction, similar to the one formed by an accelerating piston in a closed cylinder as described above. This P—V curve is rather smooth. However, when we deal with a detonation wave, which is nothing but a shock wave sustained by an explosive reaction, it becomes more complex, as will be seen in the following subsection.

3.4.2 Detonation Wave

The study of detonation was first performed in the laboratory in 1881 by detonating an explosive mixture of gases by igniting it in a long uniform tube at one end. The initial combustion wave, which was subsonic, was found to accelerate rapidly to a high constant speed, which we now know as the detonation velocity, or VOD.

Table 3.2: Detonation velocity of some common explosives.

Name of Explosive	Molecular Formula	Velocity of detonation (VOD) (km/s)	Name of Explosive	Molecular Formula	VOD (km/s)
TNT	$C_7H_5N_3O_6$	6.9	CL-20	$C_6H_6N_{12}O_{12}$	9.1
RDX	$C_3H_6N_6O_6$	8.44	PETN	$C_5H_8N_6O_{18}$	8.4
HMX	$C_4H_8N_8O_8$	9.1	TATB	$C_6H_6N_6O_6$	7.35
NG	$C_3H_5N_3O_9$	7.6	NC (dry)	$C_{12}H_{14}N_6O_{22}$	7.3
Tetryl	$C_7H_5N_5O_8$	7.57	HNS	$C_{14}H_6N_6O_{12}$	7.12

TNT: trinitrotoluene; RDX: research and development explosive; HMX: high melting explosive; NG: nitroglycerine; Tetryl: 2,4,6-trinitrophenyl-methyl-nitramine; CL-20: China Lake-20; TATB: triamino trinitro benzene; HNS: hexanitrostilbene; PETN (pentaerythritol tetranitrate); NC: nitrocellulose. For their molecular structure refer to fig 1.4, table 4.4 and table 11.1

The value of the VOD was found to depend mainly on the composition of the explosive mixture and not on the tube material, tube diameter (beyond certain minimum), and method of initiation. Typical detonation velocities, temperatures, and pressures in gas mixtures are in the range of 2000 m/s, 3000 K, and 2 MPa (20 bar), respectively. The detonation velocity of a few common explosives is given in Table 3.2.

As mentioned previously, for a shock wave without a chemical reaction, the Hugoniot curve passes smoothly from its initial state (P_o, V_o) (Figure 3.5). In case of detonation, it is not so. Detonation can be thought of as a two-step process in which a chemical reaction releases energy in a constant-volume explosion (point X) and the reaction products are then shock-compressed to some final state (point B). The velocity of the final state is proportional to the slope of the line passing through the initial and final states (Rayleigh line). This can be obtained by eliminating U_o between Eqns (3.1) and (3.2) as follows:

$$U_s = V_o \left[\frac{P - P_o^{1/2}}{V_o - V} \right] \tag{3.4}$$

The RH equations cannot by themselves predict which of the Rayleigh lines (OA or OB) corresponds to the unique detonation velocity. Chapman–Jouguet (C–J) theory makes an assumption that

$$D = C + U_p \tag{3.5}$$

where D = the velocity of the detonation front,

C = the velocity of sound in the medium, and
U_p = the velocity of the detonation products

This is given by drawing a tangent from the initial state (P_o, V_o) to the Hugoniot curve (OA). Point A is called the "C–J point." The application of C–J theory to solid explosives is more complex. Here, the products form a very dense gas for which the P-V-E relationship is not well known; hence, the computed properties are less accurately predicted.

3.5 Detonation Theory

The development of a proper detonation theory was a complex task because the very process of detonation itself is complex. It has to deal with the chemistry of a very fast, exothermic reaction; changes of mass, momentum, and energy during the reaction from reactant to products; very high pressure, temperature, and density changes; the nonideal behavior of product gases at high pressures; and so on. In the later part of 19th century and early part of 20th century, Chapman, Hugoniot and Joguet studied the thermodynamics of shock waves and its extension to reactive systems. This led to the development of the so-called "hydrodynamic theory of detonation." The mathematical treatment of this theory is beyond the scope of this book, and those interested might refer to some of the books given in the Reference section of this chapter. The author aims to highlight only the salient points of this theory to help the reader understand the concepts and approach of the theory.

During a detonation, an explosive chemical reaction is initiated immediately in the wave front because of the drastic temperature and pressure conditions. Apart from these two parameters, there is a significant difference between the undetonated explosive and the molecules in the shock zone with respect to density (ρ), specific volume (V; i.e., volume occupied by 1 g of substance – inverse of density), internal energy (E), and the velocity of sound in the medium (c). Figure 3.6 shows that there is a sudden, discontinuous jump in all of these parameters at the interface between the shock zone (subscript 1) and undetonated explosive (subscript o).

The sudden discontinuity in such parameters was mathematically treated using the following laws and conditions:

1. Law of conservation of mass (before and after explosion),
2. Law of conservation of energy (internal energy),
3. Law of conservation of momentum,
4. The equation of state (for gases), and
5. It is also assumed that the velocity of the detonation wave is equal to the sum of the velocity of sound in the medium and velocity of the products.

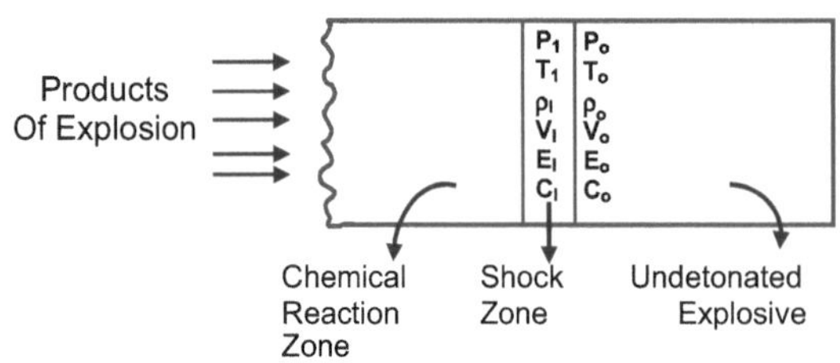

Figure 3.6
Discontinuity between Shock Zone and Undetonated Explosive.

The following are the salient points worth mentioning as the outcome of the hydrodynamic theory of detonation:

1. The relationship between detonation velocity, detonation pressure, and density.

It can be shown (see Figure 3.6) that

$$P_1 = \rho_0 D U_p, \quad (3.6)$$

where P_1, ρ_0, D, and U_p represent, respectively, the detonation pressure, the density of the undetonated explosive, the VOD, and the VOD products. Combining Eqns (3.4) and (3.5) and using the adiabatic condition (PV^γ = constant) and equation of state, one finds that U_p is related to D as

$$U_p = D/\gamma + 1 \quad (3.7)$$

where γ is the ratio of specific heats of gaseous products. Substituting Eqn (3.7) in Eqn (3.6),

$$P_1 = \rho_0 D.(D/\gamma + 1) \quad (3.8)$$

Under the detonation conditions of high temperatures and pressures in the shock zone, the value of γ of gases is approximately 3 and Eqn (3.8) becomes

$$P_1 = \frac{\rho_0 D^2}{4} \quad (3.9)$$

Hence, any increase in the density of an explosive exponentially boosts the detonation pressure, showing the importance of the density of high explosives.

2. Hugoniot curve and C–J pressure.

We have seen in Figure 3.5 that the Hugoniot curve describes the locus of all P–V states attained by shock wave compression. Some of the interesting points that can be noted from this curve can be summarized as follows:

- Zone 1 (bottom right quadrant with respect to initial state (P_o, V_o) is the deflagration zone, where $V > V_o$ and $P < P_o$ i.e., the deflagration products expand rapidly and there is no compression).
- Zone 4 is the detonation zone, where $P > P_o$ and $V_o > V$ (see Eqn (3.4)).
- As mentioned earlier, point A is called the C–J point. It is at this point where the detonation is stable. Above this point (e.g., at point B), the rarefaction wave catches up with the detonation wave; therefore, the detonation dies out. On the other hand, at point A, the detonation wave is constantly sustained by the chemical energy and products of explosion. At this steady state, the detonation has a constant intensity

and constant velocity. Under this condition, the VOD becomes equal to the sum of the velocity of sound through the medium and the velocity of the detonation products. This can be written as

$$D_{CJ} = C_{CJ} + U_{CJ}$$

as described earlier (Eqn (3.5))

- Zone 2 (where $V_o > V$ and $P_o > P$) and Zone 3 (where $V_o < V$ and $P_o < P$) do not have any physical significance because the substitution of these values results in imaginary values for shock wave velocity U_s (in case of detonation, we call it D). The values of VOD and the detonation pressure (P_d) of explosives vary anywhere from 1500 to approximately 9500 m/s and from 2 to 50 GPa, respectively. (Note: 1 GPa = 1 gigapascal = 10^9 Pa; 10^5 Pa = 1 bar ≈ 1 atm pressure. From these relations, it is seen that 1 GPa = 10^4 bar = 10 kilobar = 10 kbar. Both of the units of gigapascals and kilobars are used while quoting the values of P_d of explosives. e.g., 40 GPa or 400 kbar).

3.6 Theoretical Estimation: VOD and P_d

Several attempts have been made over the last many decades to theoretically predict the VOD and P_d of explosives. Four popular methods of VOD calculation are briefly mentioned in the following subsections.

3.6.1 Kamlet–Jacob Method

The Kamlet–Jacob (KJ method) method, developed by M. J. Kamlet and S. J. Jacobs of the Naval Ordnance Laboratory (United States) assumes that during detonation of a CHNO explosive, carbon dioxide (CO_2) and carbon (C) are preferentially formed rather than carbon monoxide (CO). The detonation equation was derived accordingly, and the same is given in Eqn (3.10):

$$D = A\left[NM^{1/2}(-\Delta H_d)^{1/2}\right]^{1/2}\left(1 + B\rho_o\right) \tag{3.10}$$

where D = VOD, A = a constant having a value of 1.01,

N = the number of moles of gases evolved per gram of the explosive,
M = the average molecular weight of the gases,
B = a constant having a value of approximately 1.30,
ρ_o = the density of the unreacted explosive in grams per cubic centimeter, and
ΔH_d = the heat of detonation (explosion) in calories per gram.

Example: Let us calculate the VOD of TNT, having molecular formula $C_7H_5N_3O_6$, at its density of 1.64 g/cc. The heat of detonation is 1090 cal/g.

Step 1: Write the detonation equation.

$$C_7H_5N_3O_6 \rightarrow 1.5N_{2(g)} + 2.5H_2O_{(g)} + 1.75CO_{2(g)} + 5.25C$$

(Note: CO is not written as a product)

Step 2: Calculate the number of moles of gases formed per gram of TNT (molecular weight of TNT = 227).

$$N = \frac{1.5 + 2.5 + 1.75}{227} = 0.02532$$

Step 3: Calculate the average molecular weight of the gases formed.

$$M = \frac{(1.5 \times 28) + (2.5 \times 18) + (1.75 \times 44)}{5.75} = 28.51$$

Using the above formula of the KJ method.

$$D = 1.01\left[(0.02532)(28.51)^{1/2}(1090)^{1/2}\right]^{1/2}\left[1 + (1.30 \times 1.64)\right]$$

$= \mathbf{6680\ m/s}$ which fairly agrees with the experimental value of 6930 m/s.

Detonation pressure (P_1):

From Eqn (3.9),

$$P_1 = \frac{\rho_0 D^2}{4}$$

Converting the values of ρ_0 (density of explosive) into SI units,

$$\rho = 1.64\ g/cc = 1.64 \times 10^3\ Kg/m^3$$
$$D = 6680\ m/s = 6.68 \times 10^3\ m/s$$

Substituting these values in Eqn (3.9),

$$P_d = 1.64 \times 10^3 \times (6.68 \times 10^3)^2$$
$$= 18.3 \times 10^9 Pa$$
$$= 18.3\ GPa$$

(Experimental value is ~21.0 GPa).

3.6.1.1 CO first or CO_2 first?

While writing the chemical equation for detonation, the explosives chemist faces this quandary. In the case of underoxidized products resulting from low or highly negatively oxygen-balanced explosives, the CO-first approach gives higher values of n and lower values of M. It also appears from limited experimental data (cf., American Institute of Physics Handbook, 2nd ed., McGraw Hill Publishers, New York, 1963) that equilibrium shifts toward CO at a lower density of loading of explosives and toward CO_2 when the density of loading is higher.

As a sort of a thumb rule, we can write down the CO-first equation when the OB of the explosive is low or negative and/or when the loading density of the explosive is low. The converse is true for the CO_2-first equation.

3.6.2 Becker-Kistiakowsky-Wilson Method

We have learned that any gas that obeys the universal gas equation, $PV = RT$ (or $PV = nRT$ for n moles of a gas) is known as an ideal gas. However, all gases are nonideal and they deviate from the expected ideal behavior more and more at higher pressures and lower temperatures because of higher intermolecular attraction and the higher percentage of volume the molecules themselves occupy in a container. The well-known van der Waal's equation, $(P + a/V^2)(V - b) = RT$ (for 1 mol of a gas), overcame this problem but only to a limited extent.

The term a/V^2 compensates for the less pressure experienced by the walls of the container of the gases due to intermolecular attraction. The term b is the covolume that takes into account the volume occupied by the molecules themselves.

The situation becomes much worse during the process of detonation because in the detonation zone the pressure of the product gases is extremely high. The detonation pressure of explosives varies anywhere between 2 and 50 GPa (i.e., on the order of 10^5 atm). During the last 50 years, different groups of authors have attempted to evolve different equations of state that take into account of the nonideal behavior of gases at such high pressures, but again with limited success. All of these methods use model equations that do not quite satisfactorily yield the condition of highly dense and heated detonation products. This includes the Becker-Kistiakowsky-Wilson (BKW) method, which needs five parameters: pressure (P), temperature (T), internal energy (E), density (ρ), and detonation velocity (D). It also needs two separate sets of data for calculations— one set for explosives with negative OB and the other set for those with positive OB.

3.6.2.1 BKW Method

This scheme utilizes thermodynamic and hydrodynamic properties to solve a set of equations. The BKW equation of state is given as

$$P = nRT\rho\left(1 + xe^{\beta x}\right)$$

where ρ = density (inverse of specific volume),

$x = b\rho\kappa T^{-\alpha}$ (b is the covolume),
$\beta = 0.3$, $\alpha = 0.25$, and $\kappa \approx$ unity.

The above equation makes allowance for the compressibility of the molecules of the product gases at very high pressures in the detonation front. The values given for α, β above are found to be an optimal fit on the basis of experimental results. On the basis of this, several computer programs (RUBYCODE, STRETCH BKW, TIGERCODE, LOTUSES) have been worked out for the calculation of VOD, P_d, and the temperature of detonation.

Because of the iterative nature and arduous calculations involved, this scheme is performed by a computer program. The equation has four arbitrary constants—α, β, θ, and κ—that require calibration to suit any particular type of explosive. An attempt has been made to obtain a unique set to satisfy many explosive compositions.

3.6.3 Rothestein and Petersen Method

One method that relies only on the chemical structure of the explosive molecule is by Rothestein and Petersen (1979 and 1981). It yields values for VOD at the theoretical maximum density. A simple, empirical linear relationship between detonation velocity at theoretical maximum density and factor *F*, which is dependent solely on chemical composition and structure, is postulated for a gamut of ideal CHNO-type explosives by L. R. Rothstein and R. Petersen. The factor *F* is expressed as

$$F = 100x\frac{\mathrm{nO} + \mathrm{nN} - \frac{\mathrm{nH}}{2\mathrm{nO}} + \frac{A}{3} - \frac{\mathrm{n}B}{1.75} - \frac{\mathrm{n}C}{2.5} - \frac{\mathrm{n}D}{4} - \frac{\mathrm{n}E}{5}}{\mathrm{MW}} - G$$

$$D = \frac{F - 0.26}{0.55}$$

where *nH*, nN, and nO are the number of hydrogen, nitrogen, and oxygen atoms in a molecule;

n*B* is the number of oxygen atoms in excess of those already available to form CO_2 and H_2O;
n*C* is the number of oxygen atoms doubly bonded to carbon as in a carbonyl group;

nD is the number of oxygen atoms (other than those in $-O\text{-}NO_2$ group) singly bonded to carbon;
nE is the number of nitro groups existing either as in a nitrate ester configuration or as a nitric acid salt such as hydrazine mononitrate;
$A = 1$ if the compound is aromatic, otherwise $A = 0$;
$G = 0.4$ for a liquid explosive and $G = 0$ for a solid explosive;
$F =$ factor; and
$D =$ the detonation velocity in kilometers per second.

To achieve the maximum VOD for a homogeneous explosive, it is necessary to consolidate the explosive composition to its maximum density.

3.6.3.1 *Illustration*

Let us examine detonation velocity of NG as shown below.

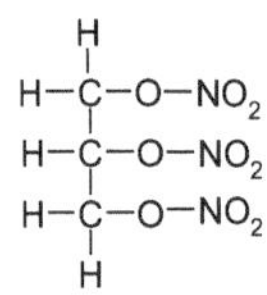

Nitroglycerine

The empirical formula of NG is $C_3H_5N_3O_9$. The explosion reaction of NG is
$C_3H_5N_3O_9 \rightarrow 3CO_2 + 2\frac{1}{2}H_2O + 1\frac{1}{2}N_2 + \frac{1}{4}O_2$

$A = 0$ because NG is nonaromatic,
$G = 0.4$ because NG is a liquid,
$nO = 9$ because the number of oxygen atoms in the NG molecule is nine,
$nN = 3$ because the number of nitrogen atoms in NG is three,
$nH = 5$ because the number of hydrogen atoms in NG is five,
$nB = 0.5$ because nine oxygen atoms are available (2.5 of these are required to form 2.5 mol of H_2O from the five hydrogen atoms and six of the oxygen atoms are needed to form 3 mol of CO_2 from the three carbon atoms, leaving 0.5 oxygen atom (or 0.25 oxygen molecule)),
$nC = 0$ because no oxygen atoms are double bonded to carbon in the NG molecule,
$nD = 0$ because all oxygen atoms in the molecule belong only to $-O\text{-}NO_2$ groups,
$nE = 3$ because there are three nitrate ester groups, and
$MW = 227.1$ (the molecular weight of NG).

Armed with these variables, we can calculate the value of detonation factor F:

$$F = 100\left(\frac{9 + 3 + 0 - \frac{5-0}{2\,x\,9} + \frac{0}{3} - \frac{0.5}{1.75} - \frac{0}{2.5} - \frac{0}{4} - \frac{3}{5}}{227.1}\right) - 0.4 = 4.372$$

$$D' = \frac{4.372 - 0.26}{0.55} = 7.48 \text{ km/s}$$

From the literature, the detonation velocity of NG is found to be 7.60 m/s. The error of the estimation in this example is 100 (7.48−7.60)/7.60 = −1.6%.

3.6.4 Stine Method

A relatively accurate method of estimating detonation velocities for CHNO explosives (Stine, 1990) is based on using the atomic composition of either a pure or mixed explosive, along with the explosive's density and heat of formation. In this method, the explosive composition is defined as $C_aH_bN_cO_d$, where *a*, *b*, *c*, and *d* are atomic fractions (i.e., *a* is the number of carbon atoms in the molecular formula divided by the total number of all atoms in the molecular formula, etc.). The equation is given by

$$D = 3.69 + (-13.85a + 3.95b + 37.74c + 68.11d + 0.6917\Delta H_f)\left(\frac{\rho}{M}\right)$$

where ρ is the initial explosive density (g/cm^3), ΔH_f is the heat of formation of the explosive (kcal/mol), and M is the molecular weight of the explosive.

3.7 Deflagration-to-Detonation Transition

Let us consider the deflagration of a propellant stick. For a given composition, at a given ambient pressure, the propellant burns at a fixed rate (linear burn rate, *r*). We mentioned two important equations in Sections 3.2 and 3.3: $r = bP^n$ (exponential dependence of *r* on P) and $\dot{m} = rA\rho$ (relationship between the linear burn rate (*r*) and $\dot{m}$ the mass burn rate ($\dot{m}$)).

As long as the rate at which the deflagration products evolved (i.e., $\dot{m}$) is equal to or less than the rate at which they are removed from the scene (e.g., m_r), one does not expect any accumulation of product gases around the burning propellant resulting in increased pressure around it. However, if $\dot{m} > m_r$, it results in an increase in pressure around the burning propellant. Higher pressure leads to a higher value of *r*. Higher *r* means still a higher rate of buildup of pressure. It becomes a superfast, vicious cycle between pressure increase and $\dot{m}$ increase until at one stage the value of *r* exceeds the sonic velocity of the medium (burning propellant). Once *r* exceeds the sonic velocity, as we have seen in Section 3.5, it leads to the formation of the vertical fronted (shock) detonation wave. This is known as the DDT, a very important phenomenon in the field of explosives.

3.7.1 When Can DDT Occur?

1. When there is a high degree of confinement experienced by the deflagrating material.
2. If the deflagrating explosive is initiated by a high-intensity shock wave.

3. In the presence of a large degree of porosity in deflagrating material (which means a very large surface area of exposure—remember, $\dot{m} = rA\rho$, where A is the area exposed for burning).
4. In large explosive charges in which the bulk of the explosive itself provides necessary confinement—particularly when they are in the form of finely divided material. (Granular TNT initiated with black powder burns quickly if the TNT is spread in thin layers on the ground ($\dot{m} < m_r$). It is bound to detonate if piled up in a large mound ($\dot{m} > m_r$)).

Therefore, during disposal of waste explosives or propellants, one has to ensure that the material is spread into thin layers to avoid the DDT phenomenon. Some of the disastrous explosions involving some seemingly innocuous materials are known to have been caused by DDT. DDT studies are essential to avoid unwanted and catastrophic detonations. During the development of new propellant compositions and scaling up of the processing of explosives, DDT studies should be performed. For example, one cannot afford the loss of a costly gun barrel if the newly developed gun propellant undergoes DDT. Likewise, scaling up the production of explosives without DDT trials may destroy the production plant because there exists a possibility of detonation because of the mass effect.

Suggested Reading

[1] S.M. Kaye (Ed.), Encyclopaedia of Explosives and Related Items, vols 1−10, US Army, Armament R&D Command, NJ, 1983.
[2] J. Taylor, Detonation in Condensed Explosives, Clarendon Press, Oxford, 1952.
[3] S.S. Penner, B.P. Mullins, Explosions, Detonations, Flammability and Ignition, Pergamon Press, London, New York, 1959.
[4] C.H. Johnson, P.A. Persson, Detonics of High Explosives, Academic Press, London, New York, 1970.
[5] W. Fickett, W.C. Davis, Detonation, University of California Press, Berkeley, 1979.
[6] R. Cheret, Detonation of Condensed Explosives, Springer Verlag, New York, Berlin, 1993.
[7] Service Textbook of Explosives, Min of Defence Publication, UK, 1972.
[8] C.S. Robinson, Explosions, Their Anatomy and Destructiveness, McGraw-Hill Book Co. Inc, New York, London, 1944.
[9] P.W. Cooper, Explosives Engineering, VCH Publishers, Inc, USA, 1996.
[10] B. Zeldovich Ia, A.S. Kompaneets, Theory of Detonation, Academic Press, New York, USA, 1960.
[11] A. Bailey, S.G. Murray, Explosives, propellants & pyrotechnics, in: Land Warfare: Brassey's New Battlefield Weapon Systems & Technology Series, vol. 2, Royal Military College of Science, Shrivenham, UK, 1989.
[12] L.R. Rothstein, R. Petersen, Predicting high explosive detonation velocities from their composition and structure, Propellants Explos. 4 (1979) 56−60.
[13] L.R. Rothstein, Predicting high explosive detonation velocities from their composition and structure (II), Propellants Explos. 6 (1981) 91−93.
[14] J.A. Zukas, W.P. Walters (Eds.), Explosive effects and applications, Springer-Verlag, New York, USA, 1997.

Questions

1. How do you define explosion and how can it be classified?
2. All explosions result in gaseous products with increased volume. Is this statement always true? If not, then give some examples that are contrary to the above statement.
3. What are the differences between deflagration and detonation?
4. Why does an ordinary, innocuous fuel become dangerous when it is finely dispersed in air?
5. Ten kilograms of a propellant are to be disposed of by open burning after breaking into cubical pieces, each piece having a length of 1 cm. If the density of the propellant is 1.5 g/cc and its linear burning rate is 2 mm/s at atmospheric pressure, then what will be the initial mass burning rate when the entire lot of propellant is simultaneously ignited? (Answer: 12,000 gs^{-1}).
6. Why does a shock wave assume a plane wave front?
7. What are the different zones identifiable when a column of solid explosive undergoes detonation?
8. Why do detonation products move along the direction of the propagation of detonation?
9. β-HMX has a detonation velocity of 9100 m/s and density of 1.96 g/cc. What is the expected detonation pressure (in GPa) when β-HMX detonates? (Answer: 40.58 GPa).
10. What conditions can favor DDT?
11. Name some computer programs developed for the calculation of VOD and detonation pressure.
12. Name some methods for the theoretical estimation of VOD.
13. What is the difference between linear burning and mass burning?
14. Give the relationship among the detonation velocity, detonation pressure, and density of an explosive.
15. What is the significance of the Hugoniot curve?

CHAPTER 4

HEMs: The Facet of Explosive Performance

4.1 Why Do Explosives Explode?

4.1.1 Spontaneity of Explosive Reactions

Why do explosives explode? We have seen in Chapter 2 that all explosives are basically metastable and they just wait for a trigger energy. Their explosive reactions—whether the explosive has a positive or negative heat of formation—are highly exothermic and evolve large volumes of gaseous products. Once the trigger energy is given, an explosive reaction proceeds spontaneously and no one can just stop it.

What decides spontaneity? All changes in nature, whether physical changes or chemical changes, are governed by two seemingly opposing forces, viz:

1. Tendency for minimum energy
2. Tendency for maximum randomness (freedom or disorder)

A boulder on the precipice of a hill tends to come down to minimize its energy content. It is too tense to have so much potential energy! A compressed gas in a cylinder tends to come out through the nozzle so that its molecules can go away from each other, enjoying complete freedom or randomness. They feel so congested and crowded within the cylinder! In thermodynamic parlance, the energy is referred by "H" ***(heat content)*** and the randomness or degree of disorder by "S" ***(entropy)***. However, often these two tendencies may oppose each other. To decide whether a process can take place spontaneously or not, both these tendencies are simultaneously taken into account at a given temperature "T", and a parameter called "***Gibb's free energy***" is introduced. G is defined as:

$$G = H - TS$$

At a given (constant) temperature, the changes in H, S, and G are then:

$$\Delta G = \Delta H - T\Delta S$$
$$-\Delta G = -\Delta H + T\Delta S$$

Ultimate criterion Tendency for spontaneity minimum energy ($-\Delta H$) maximum randomness ($+\Delta S$)

Demystifying Explosives: Concepts in High Energy Materials. http://dx.doi.org/10.1016/B978-0-12-801576-6.00004-5

Any system in nature tries to undergo a change resulting either in $-\Delta H$ (release of energy) or $+\Delta S$ (increase in disorder), or both. However, what happens if a change has opposing tendencies like ($-\Delta H$ and $-\Delta S$) or ($+\Delta H$ and $+\Delta S$)? Substitute these values in the above equations and see the sign of ΔG. If it is negative, the process takes place spontaneously. If it is positive, it does not. It all depends on the relative values of ΔH and ΔS. Let us take two examples of such opposing tendencies resulting in spontaneous changes:

1. Endothermic dissolution ($+\Delta H$) of salts in water:

The freedom of solvated ions in solution ($+\Delta S$) as compared to the "confinement" of the cations and anions in crystal lattice overshadows the endothermic ($+\Delta H$) effect to achieve a negative value for ΔG.

2. Polymerization reaction:

When a large number of monomer molecules result in a fewer number of macromolecules, their overall randomness ($+\Delta S$) is reduced. However, the high exothermicity ($-\Delta H$) of polymerization reaction overshadows this effect, resulting in a negative value for ΔG.

An explosive reaction is both highly exothermic ($-\Delta H$) and involves the production of a large number of gaseous products ($+\Delta S$), thus involving a large decrease in the values of free energy and making the process of explosion highly spontaneous.

4.1.2 The Kinetic Aspect of Explosive Reactions

For an explosion to occur, the reaction must be exothermic, large amounts of gases must be produced by the chemical reaction and vaporization of products, and the reaction must propagate very fast. If a chemical change is accompanied by a large decrease in free energy, i.e., $-\Delta G$, does it mean that it can take place easily? Not necessarily. For example, a piece of coal burns in oxygen to form CO_2 and water, and this exothermic burning should involve a large decrease in free energy. Even such "spontaneous" chemical changes need to be given an initial "**activation energy**" (E_a), to "prepare" the reactants for instant chemical change. We have seen in Chapter 2 that even the so-called metastable explosives need to be given a "trigger energy" or activation energy to undergo explosive decomposition.

If this barrier of E_a does not exist, nothing will be stable on this earth, whether an explosive or a piece of coal! The difference may be that the value of E_a in the case of wood may be far more than that of an explosive. The less stable (or "metastable") explosive molecule X has all its bonds intact (See Figure 4.1) in its ground state. Once it is given the trigger or activation energy (by shock, heat, or impact), the molecule goes to the excited or activated state (X*) where certain bonds are preferentially broken (like the

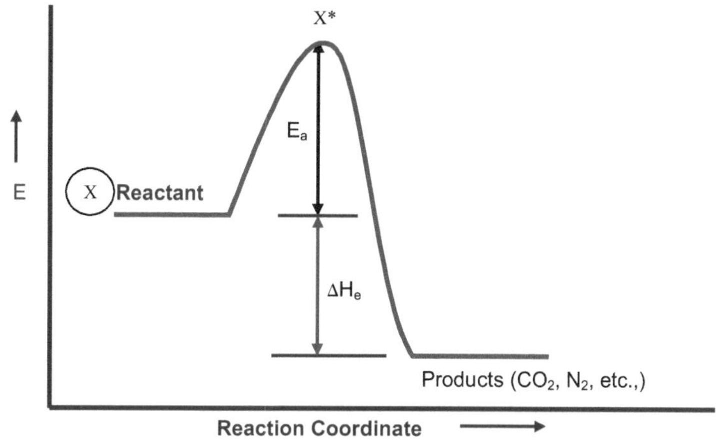

Figure 4.1
Activation Energy: A Need for a Chemical Reaction.

breakage of C—ONO_2 bond in nitric esters like NG). Having absorbed so much energy and having been mutilated like this, the molecule will no longer remain in that position but will immediately go to the lower energy state by degrading into stable products like CO_2, CO, N_2, etc., releasing large amount of heat.

The rate of such a reaction will depend mainly on two parameters viz. (1) the value of activation energy; and E_a (2) temperature of the reaction condition, as given by the Arrhenius equation for any chemical reaction:

$$k = Ae^{-Ea/RT}$$

where k = rate constant of the reaction

A = temperature-independent "Arrhenius constant" (or pre-exponential factor).

This equation shows that when temperature T increases or E_a decreases, the rate of the reaction increases exponentially, i.e., much faster than a linear relation (See Figure 4.2). If we take the logarithmic form of the Arrhenius equation, we get:

$$\log k = \log A - \frac{E_a}{RT}$$

A plot of log k against 1/T (Figure 4.3) gives, therefore, a straight line whose slope gives the value of $-E_a/R$, from which we can calculate the value of E_a.

E_a is an important parameter for an explosive. The lower the value of E_a, the higher is the sensitivity of the explosive for initiation.

Section 4.1.1 discusses the *thermodynamic* nature of explosives that have a huge "free energy advantage," whereas Section 4.1.2 talks about the *kinetic* aspect of explosives, i.e.,

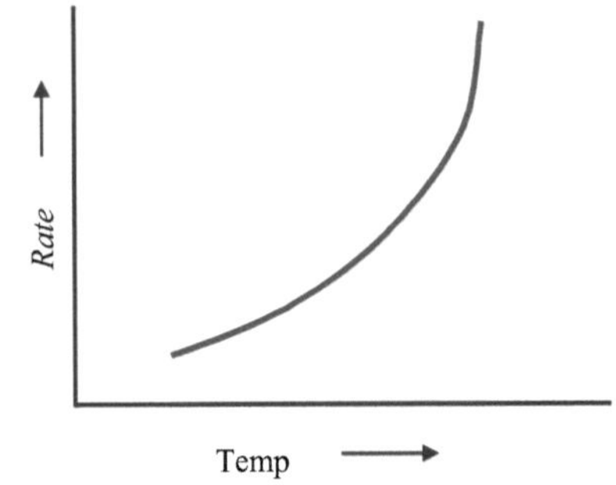

Figure 4.2
Exponential Increase of the Rate of a Reaction with Respect to Temperature.

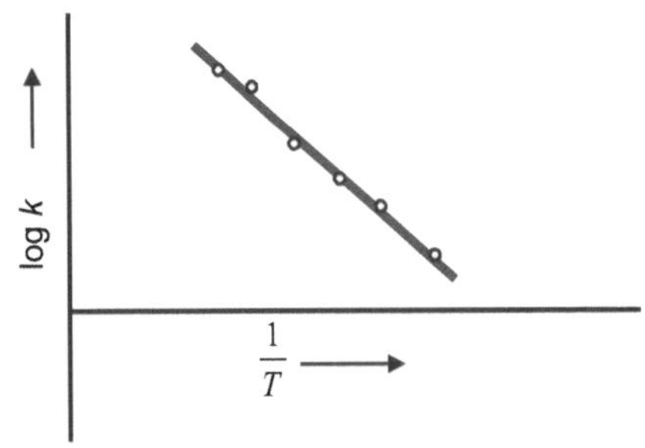

Figure 4.3
Plot of log k against 1/T.

how easily the explosive decomposition can take place with respect to the values of their activation energy and temperature.

4.1.3 Molecular Structure and Explosive Properties

The relationship between the molecular structure and the explosive property of a compound has been a matter of interest for quite some time. By the end of the nineteenth century, vant Hoff observed that nitration of certain compounds imparted explosive properties to the nitrated products. An increase in the number of nitro groups in a molecule caused an increase in the explosive properties. He stated: "It is atomic bonds of specific nature which generate the explosive character". In 1953, Plets proposed the concept of "**explosophores**" and "**auxoploses**" (similar to "chromophores" and "auxochromes" in the chemistry of dyes). Explosophores are certain specific functional groups that impart explosive properties to the molecules of the explosive compounds. Some of such groups are listed in Table 4.1.

Table 4.1: Structure of some of the explosophore groups.

Nitro	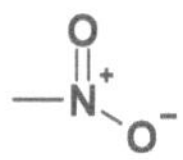	Nitrato	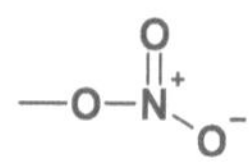
Azo	—N=N—	Azido	$—N^{-}—N^{+}\equiv N$
Peroxido	—Ö—Ö—	Perchlorato	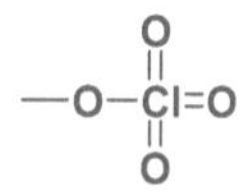

Auxoploses are functional groups that modify or increase the explosive properties in a molecule. For example, the ease of initiation of picric acid (I) as compared to trinitrobenzene (II) appears to be due to either the activation of the benzene ring by the electron-donating hydroxyl group of (I) or the stabilization of the activated intermediate product of detonation by the —OH group. The hydroxyl group in (I) acts as an auxoplose.

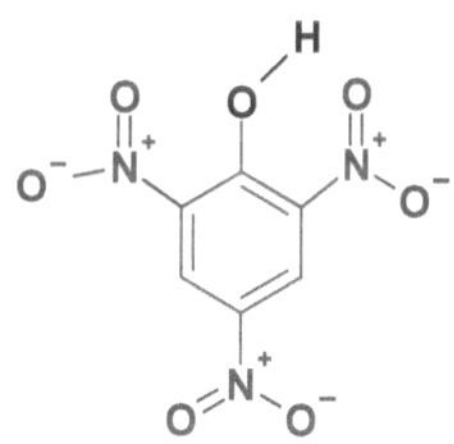

(I) Picric acid

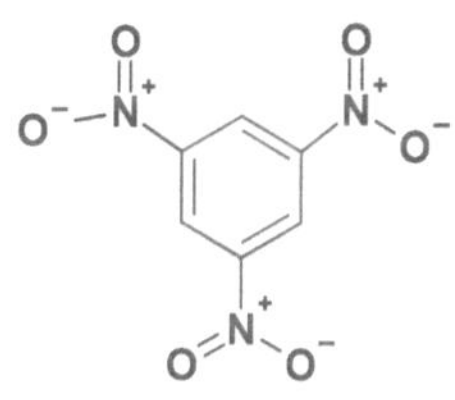

(II) Trinitrobenzene

Let us ask a basic question: Why are only certain functional groups like $-NO_2$, $-ONO_2$, $-ClO_4^-$, etc., explosive or metastable in nature? When a bond is formed between two atoms of different electro-negativities (ENs) (i.e., the tendency of an atom to attract the electrons to itself), the bond is strong; e.g., an O—H bond (wherein the EN of oxygen is higher than that of hydrogen) is strong and stable because of the dipole formed as shown below:

$$\overset{\delta-}{O} : \overset{\delta+}{H} \quad \text{or} \quad \overset{\delta-}{O} - \overset{\delta+}{H}$$

If both of the bonding atoms have high EN values (the atoms lying toward the right of the periodic table of elements), their competition for electrons is so much that the bond formed between them is unstable (e.g., N—O as in nitrates and Cl—O as in perchlorates). Although N and Cl are highly electronegative atoms, a large positive charge (or oxidation number) of 5+ and 7 + has been forced on these atoms in NO_3^- and ClO_4^-, respectively. Therefore, these NO_3^- and ClO_4^- groups badly look for electron donors to relieve this bonding stress. Carbon atoms within the same explosive molecule can oblige them to achieve this resulting in very stable products like CO_2, CO, and N_2. The well known Molecular Orbital Theory of bonding explains the relative instability of such groups and also groups such as azides.

$$-\overline{N}-N^{+}\equiv N$$

Closer insight into the mechanistic relationship between molecular structure and explosive properties of compounds became possible only after the development of spectroscopic techniques like X-ray photoelectron spectroscopy (XPS) and electron paramagnetic resonance (EPR), and theoretical treatments based on quantum mechanics, molecular shock dynamics, etc. The exact relationship between the molecular structure, shock sensitivity of the explosive compound, and the detonative decomposition is rather complex and is beyond the scope of this book. However, based on the extensive quantum chemical calculations and experimental techniques like High Speed Raman Spectroscopy, the following conclusions may be drawn:

1. The electronic structure of the explosive molecule plays a key role in its degree of shock sensitivity.
2. Only certain specific bonds in an explosive molecule are vulnerable to be broken by the shock energy causing explosive initiation, e.g., an $N-NO_2$ bond in a nitramine explosive like RDX.
3. The shock sensitivity appears to depend on the extent to which the polarity of these specific bonds gets reduced after they receive the shock energy. For example, why is RDX more "explosive" than nitroguanidine? To put it more scientifically, why is the shock sensitivity of RDX more than that of nitroguanidine? It has been found that the polarity of $N-NO_2$ bond is reduced by 55 and 22% in RDX and nitroguanidine, respectively, after they receive shock energy. It implies that $N-NO_2$ bond becomes more non-polar in RDX. A nonpolar bond gets more easily broken than a polar bond because in the latter, the electrostatic forces between the concerned atoms resist the breakage of the bond.
4. The decomposition of the molecules of an explosive compound does not occur immediately after the passage of shock front. Behind the front, the energy transmitted by the shock wave is received preferentially by certain molecules causing excitation in their vibrational and electronic energy levels. This excitation, which is a measure of the sensitivity of the explosive, triggers the initiation. It is only after sometime, probably a few nanoseconds, and the rest of the molecular population takes part in the "cooperative decomposition" process using the energy received from the shock-initiated molecules.

4.2 Two Aspects of Explosive Performance

4.2.1 Partition of Explosive Energy

When an explosive kept inside a drilled hole of a rock detonates, the rock goes to pieces. What causes this damage? Is it the shock or the gaseous mixture produced due to explosion, or both? We can divide the effects of "explosive energy" into two components viz. shock effect and gas expansion effect as shown below.

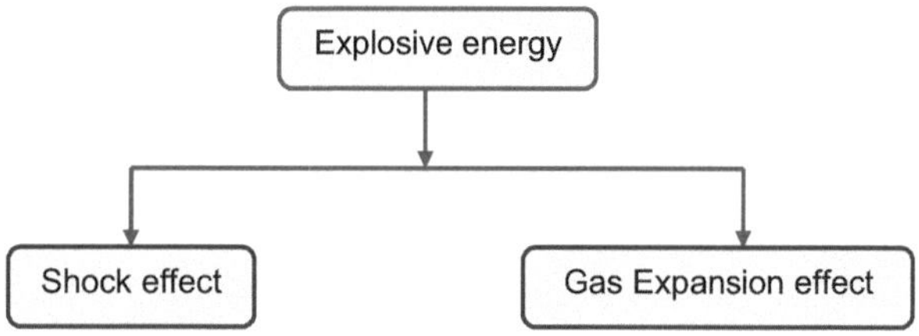

Components of energy released from an explosive

1. Shock effect: It is the effect of high pressure detonation front on the target and the effect is proportional to the Detonation Pressure (P_d) that depends on the velocity of detonation (VOD) of the explosive and its density.
2. Gas Expansion effect: It is the effect of high pressure gases produced on the target. We have seen in chapter 2 that this is equal to nRT_o where "n" is the number of moles of the gases produced per gram of the explosive and T_o is the detonation temperature. The explosive energy is therefore partitioned between shock and gas expansion.

In the above example of rock blasting, these two effects work in the following sequence:

1. The shock wave is the leader of the attack! The high pressure shock front (with pressures in the region of a few 100 kilobars) produces a highly intense compression through the rock.
2. This compression wave is followed by a rarefaction wave. During the phase of rarefaction, the pressure is below the atmospheric pressure and it creates tension (opposite of compression) through the rock.
3. The intense compression followed by tension causes plastic and elastic deformation of the material in quick succession, resulting in the cracking of the material.
4. The compressed gases waiting for such an opportunity and expand through these cracks to heave out the broken pieces.

It has been established that not more than 50% of the total explosive energy is through the shock, even in case of high VOD explosives. The share of credit between "shock" and "gas expansion" factors for a damage exercise varies depending on the application, as below:

1. In certain applications like shaped charge ammunitions (to be discussed shortly), shock plays a predominant role rather than gas expansion.
2. In certain mining applications like coal mining, gas expansion effect is more important to heave out large chunks of coal, whereas a high VOD may be disastrous.
3. In many applications, as we have seen in the rock blasting case, both these effects work hand in hand.

4.2.2 Velocity of Detonation

VOD is the rate of propagation of a detonation in an explosive. It is a major index of the explosive performance. If the density of an explosive is at its maximum value (i.e., theoretical maximum density: theoretical maximum density (TMD)) the detonation velocity is characteristic of a given explosive. VOD of an explosive, however, depends on the following major factors:

1. Loading Density (Δ)

It is defined as the ratio between the weight of the explosive and the explosion volume, i.e., the space in which the explosive is detonated. If 10 g of RDX is loaded in an enclosed space of 20 cc, the value of Δ will be equal to 10 g/20 cc = 0.5 g/cc. The higher the loading density, the higher will be the VOD for the simple reason that, a higher loading density ensures a larger quantity of the explosive per unit volume so that larger amount of evolved energy can sustain the detonation wave.

If D_1 and D_2 are the values of VOD of an explosive at Δ_1 and Δ_2, the following empirical relation mostly holds true:

$$\frac{D_1 - D_2}{\Delta_1 - \Delta_2} = 3500$$

Marshall's formula gives the following relationship between VOD, Δ, n (number of moles of gaseous products per gram of the explosive), and T_d (detonation temperature).

$$D(\text{in m/s}) = 430\,(n\,T_d)^{1/2} + 3500\,(\Delta - 1)$$

2. Charge diameter

It we detonate a cylindrical column of explosive and measure its VOD, we will find that the velocity changes if we change the diameter of the column. VOD decreases as the diameter of the column decreases. This effect is caused by more energy losses to the peripheral sides of the column. When the diameter is large, the losses are small as compared to what is produced at the wave front. However, losses become significant as the column diameter decreases.

Let us take a cylindrical column of an explosive with length L and diameter D. If E_x is the energy produced per unit volume of the explosive during detonation and E_y is the energy lost per unit area through its side surface, the percentage of loss of energy through the side of the column will be:

$$\frac{\text{the energy lost}}{\text{the energy produced}} \times 100\% = \frac{(\pi DL \times E_y)}{\frac{\pi D^2 L \times E_x}{4}} = \times 100\%$$

$$= \frac{4E_y}{DE_x} \times 100\% = \frac{k}{D};\ (k = \text{constant})$$

Since E_y and E_x are constants, this percentage loss is found to be inversely proportional to the column diameter. At lower diameters, the percentage loss of energy increases. Below a certain diameter that we call as the ***critical diameter*** of the explosive, the percentage loss is so much that the unsupported/unsustained detonation wave dies out. The critical diameter of an explosive depends on the type of explosive, its density of packing, inclusion of air bubbles, etc. It can be as small as about 0.5 mm in the case of an initiator like lead azide and as large as 100 mm in the case of less sensitive, low density ammonium nitrate.

The concept of "critical diameter" is made use of in the introduction of "detonation trap" during the manufacture of shock-sensitive NG. These traps whose diameters are less than the critical diameter of NG do not allow the shock wave to pass through them so that an accidental detonation in one part of the NG plant is not transmitted to the other parts.

3. Degree of confinement

When the degree of confinement of an explosive increases, it increases its VOD.

4. Strength of Initiation

The higher the strength of initiation of an explosive charge (e.g., by using a powerful detonator), the VOD of the explosive is higher and vice versa. Weak detonators are used to initiate industrial explosives to achieve a lower order of VOD. Measurements of VOD were earlier made by the good old "Dautriche method." Nowadays, "Pin Oscillography Technique" and "Streak Camera Techniques" are used for more accurate measurements.

4.2.3 Gas Expansion

Like the shock effect, the expansion work done by the gases is also an index of the energy of an explosive. As already mentioned in Chapter 2 (Section 2.3.9), this is called as the

"Specific energy" of the explosive. Sometimes, terms such as "strength" and "power" are also used.

Specific energy (f) is defined as the working performance per kg of an explosive. It can be calculated from the general equation of state for gases:

$$f = PV = nRT$$

If, for example, the adiabatic, isochoric flame temperature of RDX during the detonation is given as 2800 K, what would be its specific energy? One gram per mole of RDX (i.e., 222 g of RDX) explosively decomposes to give 9 mol of gaseous products (i.e., n = 9).

$$C_3H_6N_6O_6 \rightarrow 3CO + 3H_2O + 3N_2$$

Taking the value of R as 8.314 J deg^{-1} $mole^{-1}$.

$$F = 9/222 \text{ mole g}^{-1} \times 8.314 \text{ J deg}^{-1}\text{mole}^{-1} \times 2800 \text{ deg.}$$
$$= 944 \text{ J/g : Specific energy of RDX.}$$

This work performance parameter of all explosives can be experimentally determined. The old and still reliable method is by the Trauzl Lead Block Test. In this test, a known weight of the explosive is kept in a drilled hole of a solid lead block and sealed. On initiation of the explosive, the high pressure expanding gases increase the volume of the hole. The increase in volume is measured and this is a measure of "f" of the explosive that is reported in the units of cm^3/g, i.e., the volume increase per gram of the explosive used. When we plot a graph of "f" obtained in this experiment against "nRT" values of different explosives, a nearly linear plot is obtained (Figure 4.4). The slight nonlinearity of the plot is probably due to our assumption that the high pressure gases behave ideally, which actually is not true.

Conventionally, picric acid is taken as the reference explosive to compare the power or strength of the explosives. Since n ∝ V (volume of the product gases per g of the

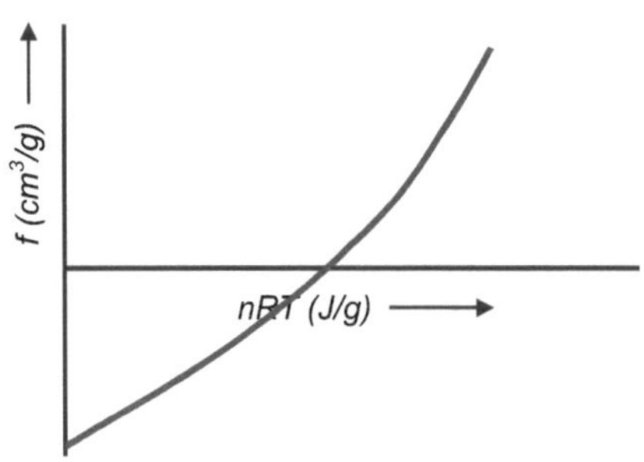

Figure 4.4
Work Performance (f) against nRT of Explosives.

Table 4.2: Power index of a few explosives computed using LOTUSES.

Explosives	Power Index (%)	Explosives	Power Index (%)
Hexanitrostilbene (HNS)	108.7	Octogen (HMX, high melting explosive)	178.33
Nitrocellulose (NC)	131.09	Pentaeryhritol tetranitrate (PETN)	177.22
Nitroglycerine (NG)	164.49	Trinitrotoluene (TNT)	103.68

explosive) and $T \propto Q$ (cal val of the explosive), we can write $nRT \propto QV$, where QV is called the ***characteristic product*** of an explosive indicating its power. The power of an explosive is compared with that of a standard explosive like picric acid and expressed as "Power Index" as shown in the following example.

If we compare RDX with the standard explosive viz. picric acid, ($Q = 1226$ cal/g, $V = 908$ cm^3/g for RDX, $Q = 896$ cal/g, $V = 780$ cm^3/g for picric acid):

$$Power\ index = \frac{Q_{RDX} \times V_{RDX}}{Q_{PA} \times V_{PA}} = \frac{1226 \times 908}{896 \times 780} = 1.59 \text{ or } 159\%$$

The power index of a few explosives was worked out using the LOTUSES software and is presented in Table 4.2, in which the volume of detonation gases were computed by Kistiakowsky–Wilson rules.

4.3 A Travel by Explosive Train

As mentioned in Chapter 1, "safety" and "reliability" are two vital requirements of any system containing explosives, including ammunitions. By "safety" we mean that the explosive should not go off when it should not go off. By "reliability" we imply that it should go off when it should go off! But explosives differ in their ***sensitivity*** to initiation by impact, friction, heat, electrical impulse, etc. as well as in their ***output*** in terms of shock intensity. To achieve our dual objective of safety of handling, transport, and storage on one hand and reliability of performance in an explosive-based system like an ammunition or an industrial explosive system on the other hand, we have to form a "train" of explosive system that contains (1) an initiating explosive of high sensitivity (HS) and low output (LO), like primary explosives; (2) an intermediary (also called booster) explosive of medium sensitivity (MS) and medium output (MO); and finally (3) the main explosive of low sensitivity (LS) and high output (HO). An example of such an explosive train is given in Figure 4.5 below:

4.3.1 *Primary Explosives:* "the engine starters in an explosive train!"

Primary explosives (or initiatory explosives) start the engine in the explosive train. They are highly sensitive to heat, friction, impact, shock, and electrostatic energy. They initiate

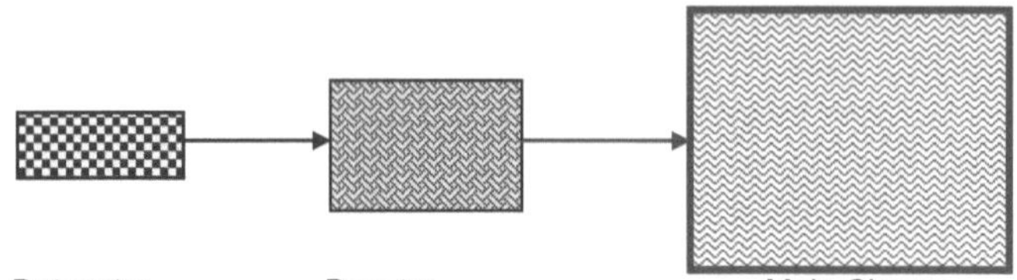

Role	Detonator (HS/LO)	Booster (MS/MO)	Main Charge (LS/HO)
Name of the Compound	Lead Azide + Lead styphnate	PETN	HMX
RIT Value*	9%	15%	35%
VOD (m/s)	5100	8300	9160

(*RIT stands for the Relative Impact Test with respect to TNT as a standard explosive. It gives the height at which a given falling weight will initiate the explosive as compared to TNT. For a given falling weight, if TNT is initiated at 100 cm height, PETN gets initiated even at 15 cm height).

Figure 4.5
An Explosive Train.

the next element or component in an explosive train (booster) that, in turn, initiates the main HE charge. Overall energy amplification from initiator to the main charge can be as high as about 10 million to one. The primary explosives are used in military detonators, commercial blasting caps, and percussion electric primers. They can be initiated electrically (e.g., "squibs"), mechanically (e.g., in stab detonators) or by shock from an exploding wire.

Only a few compounds can act as primary explosives to meet the military and industrial requirements in terms of sensitivity, output, ease of manufacture, cost, compatibility, and long-term storage stability. Mercury fulminate was the earliest one to be used, but is almost phased out now because of its poor storage stability. Some of the common primary explosives used today are lead azide, silver azide, lead styphnate, and lead dinitroresorcinate. Lead azide, although popular, has poor "flash pick-up" property and incompatibility with copper and brass used in ammunitions (due to the formation of dangerously sensitive copper azide during storage), and lead azide is normally mixed with lead styphnate to improve its flash sensitivity. Some important properties of three well-known primary explosives are given in Table 4.3.

Table 4.3: Properties of a few primary explosives.

Property	Mercury Fulminate	Lead Azide	Lead Styphnate
Molecular structure	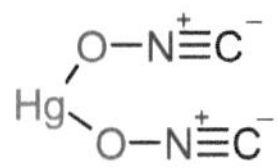	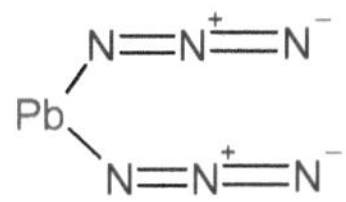	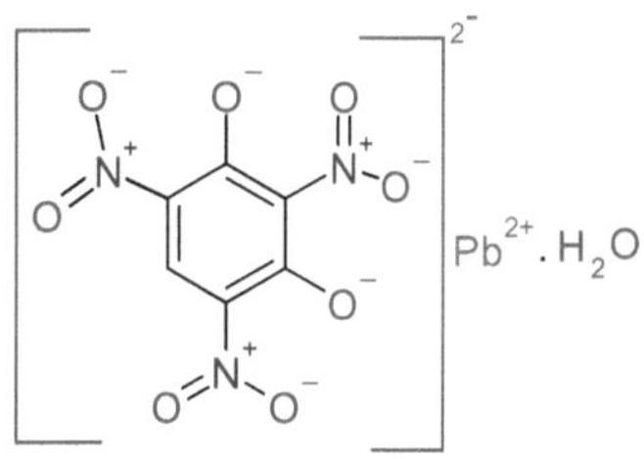
Molecular weight	284.6	291.3	468.3
ΔH_f (cal/g)	+225	+340	−451
ΔH_e (cal/g)	355	367	370
Density (g/cc)	4.2	4.8	3.0
VOD (m/s)	5400	5300	5200
Relative impact test value (%TNT)	5	11	8
Static discharge (max energy for nonignition) (J)	0.07	0.01	0.001

4.3.2 *Secondary Explosives: "the final operators of the explosive train!"*

The secondary explosives are those with low or moderate sensitivity to initiation but with higher power or output. They include boosters as well as final HE charges. Most of them come under three categories viz.

1. aliphatic nitrate esters
2. aromatic nitro compounds
3. nitramines (aliphatic/aromatic/heterocyclic)

The details of the properties, preparation, and uses of these explosives will not be discussed in this section. Only certain salient points will be mentioned.

(1) ***Aliphatic nitrate esters***: They are some of the most powerful explosives with high values of cal val, and also VOD. However, their stability is less than other classes of explosives because of the tendency of the $C-ONO_2$ bond to get hydrolyzed slowly, resulting in the formation of acids (HNO_3/HNO_2) that further catalyzes the decomposition.

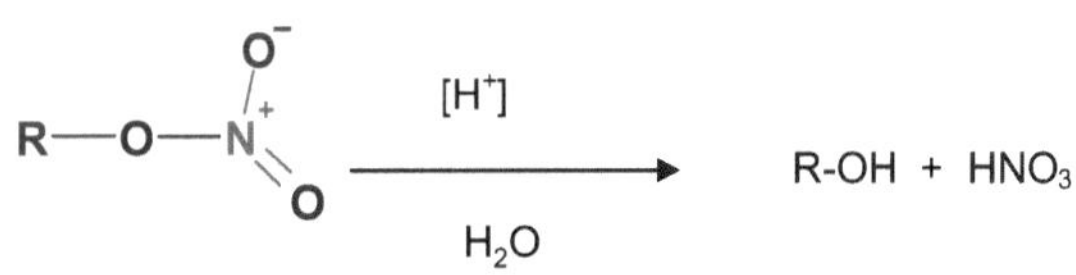

The most commonly known explosives of this category are NG, PETN, and NC. The properties of NG and PETN are given in Table 4.4.

(2) ***Aromatic nitro compounds***: Aromatic nitro compounds are more stable than nitrate esters. All aromatic molecules are ***resonance-stabilized*** and the introduction of an electron-donating group like methyl groups ("hyperconjugation") further increases the ring stability as in the case of TNT, as shown below:

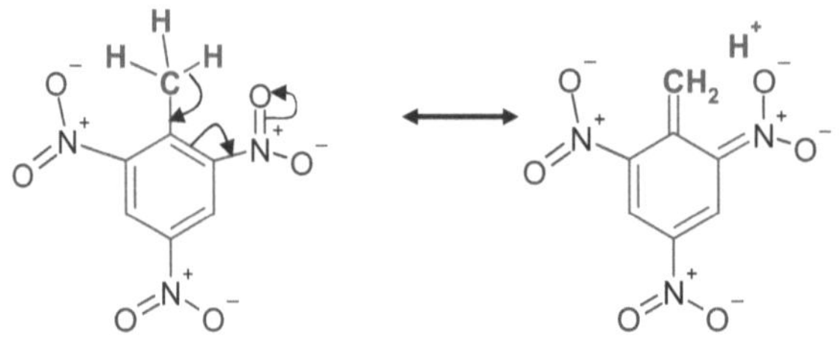

It requires more energy to destabilize such a resonance-stabilized ring, making such compounds more stable than the aliphatic explosives.

Very high thermal stability of triamino trinitrobenzene (TATB), which melts at about 350 °C, is attributed to an important additional factor viz. extensive network of hydrogen bonding as shown below:

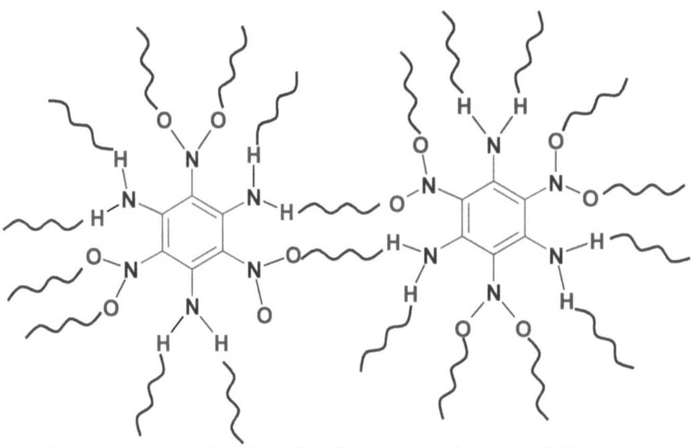

(denotes intermolecular H-bonding between H atoms of NH_2 groups and O atoms of NO_2 groups).

The important properties of two aromatic nitro compounds viz. TNT and picric acid are shown in Table 4.4.

(3) ***Nitramines***: While nitric esters represent the $O-NO_2$ explosives and aromatic nitro compounds represent $C-NO_2$ explosives, nitramines represent $N-NO_2$ explosives. They can be aliphatic like nitroguanidine, aromatic like "tetryl" (2,4,6-trinitrophenyl methyl

Table 4.4: Properties of selected secondary explosives.

Property	NG	PETN	TNT	Picric Acid	Nitro-Guanidine	RDX (Cyclonite)	HMX (Octogen)	Tetryl
Structure	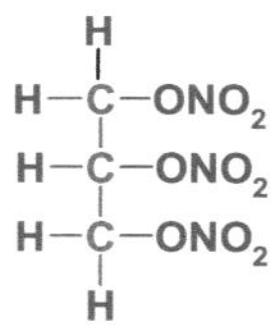	$C(CH_2ONO_2)_4$	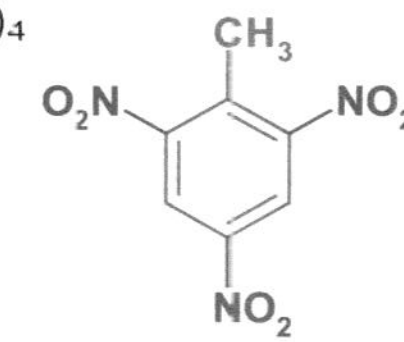	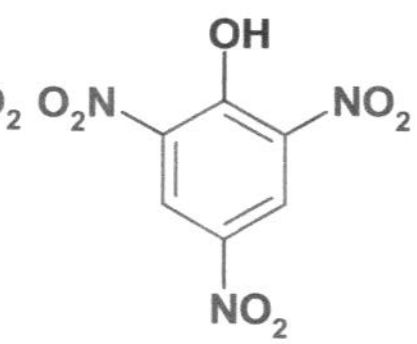	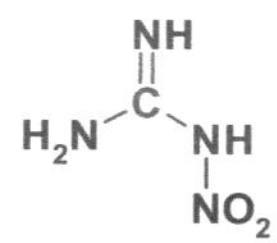	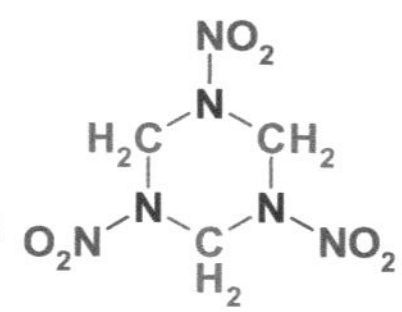	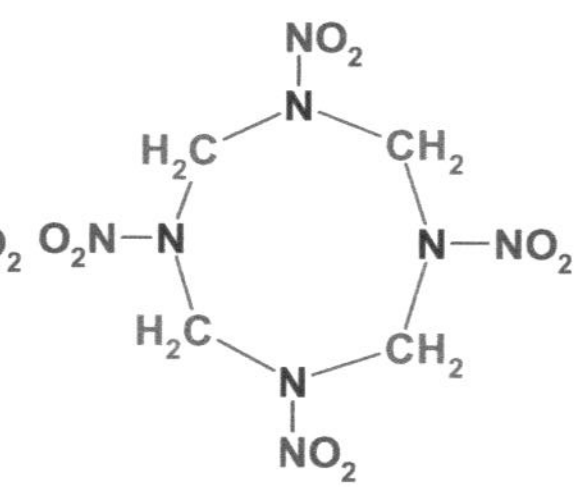	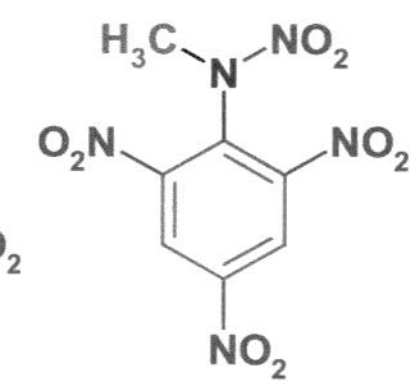
Mol. weight	227.1	316.1	227.1	229.1	104.1	222.1	296.2	287.1
ΔH_f (cal/g)	−392	−402	−62.5	−225.7	−213.5	76.1	60.4	28.1
OB (%)	3.5	−10.1	−73.9	−45.4	−30.7	−21.6	−21.6	−47.4
ΔH_e (cal/g)	1617	1529	1030	1080	769	1375	1357	1140
Density (g/cc)	1.59	1.76	1.65	1.77	1.71	1.82	1.96	1.73
m.p. (°C)	13.2	141.3	80.8	122.5	232	204	275	129.5
VOD (m/s)	7600	8400	6900	7350	8200	8750	9100	7570
Impact sensitivity (relative % to TNT)	15	20	100	100	200	35	35	50
Det. pressure (GPa)	—	32	18	26.5	27.3	33.8	39.3	26.2

nitramine), or heterocyclic like RDX (cyclic trimethylene trinitramine) or HMX (cyclic tetramethylene tetranitramine). Their properties are given in Table 4.4.

Their stability is between that of nitroaromatics and nitric esters, and they are very powerful. One reason for the higher power or high VOD of these explosives is the "energetic" (positive heat of formation) nature of $N-NO_2$ bonds. This explains the positive heat of the formation of RDX, HMX, and tetryl.

4.3.3 The Types of Explosive Trains

Today's conventional warfare cannot be imagined without the use of explosives and propellants. An explosive train is a part of any ammunition, whether it is a round fired from a pistol, large-caliber gun ammunition, hand grenade, or a guided missile. The explosive train is of two types viz.

1. Igniferous train (or "burning train")
2. Disruptive train (or "detonative train")

As mentioned earlier, in both these types of trains, the three components viz "initiator-booster-main charge" are present in the same order.

In the igniferous train, the transmission is by burning or, more correctly, deflagration, whereas in disruptive train, the transmission is by the process of detonation. In many ammunitions, both these mechanisms are operative in sequence.

Figure 4.6 schematically shows a full round of HE ammunition readily loaded in a gun for action. The full round consists of two parts. The part-1, which is shown as Explosive Train-1 (ET-1), is an igniferous train. Train ET-1 starts like this:

1. A striker pin strikes the percussion cap at the base of the cartridge and, due to high impact cum friction, the sensitive pyro composition ignites the moment the striker pin punctures the cap with force.

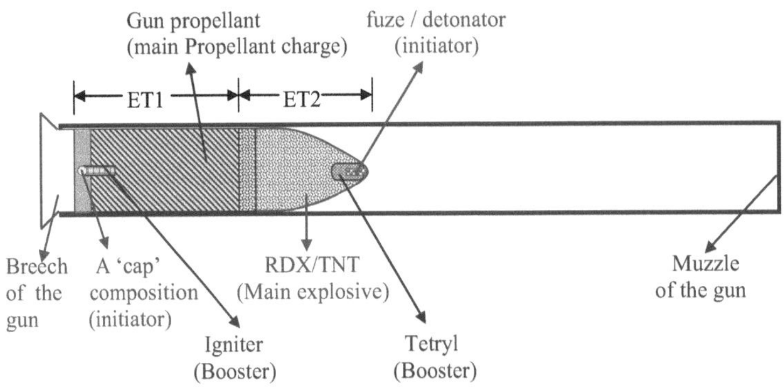

Figure 4.6
Two Types of Explosive Trains Operating in an Ammunition.

2. The initiated cap composition ignites gunpowder in the igniter housing, which burns, giving out hot particles, and the flame engulfs the entire bulk of large quantity of gun propellant.
3. The main gun propellant charge weighing a few kilograms burns in a matter of a few milliseconds, creating high pressure, high temperature gases that propel the projectile (HE shell) to a great speed.

The entire drama of the eflagration train that results in propulsion of the HE shell takes only a few milliseconds.

The second drama of "explosive train-2" (ET-2) takes place when the HE shell lands on its target (naturally, we should not have this drama enacted when it is still inside the barrel!) with the fuze getting initiated. The sequence of ET-2 will be:

1. The fuse initiates the detonator on striking the target.
2. The shock wave of the detonator is sufficient to initiate the booster pellet
3. The amplified shock wave initiates the main HE charge, wreaking havoc on the target.

4.4 Performance Parameters of Military Explosives

Interestingly, the destructive capabilities of military explosives are varied. The "**shattering effect**" of an explosive in a grenade kills or injures personnel or damages sensitive targets over a given area. The "**scabbing effect**" cuts out a chunk of armor plate of an enemy's tank and throws it right into their cabin. The overpressure of the "**blast effect**" brings down structures. The "**shaped charge effect**" causes deep penetration through heavy armor plates. The formulation of explosive composition and the overall ammunition design are so made that the final ammunition achieves one of these specific objectives. We will see in a little more detail as to how these effects operate.

4.4.1 Fragmentation

The destructive fragmentation effect of an explosive charge on its immediate vicinity is termed as "**brisance**." In a battle scenario, when a brisant explosive shatters a shell or a grenade, the high velocity fragments create extensive damage on the personnel or objects in its vicinity. What should be the shape, size, and weight of the fragments depends on the nature of the target. In case of an antipersonnel grenade, even small fragments will achieve the purpose of killing or incapacitating the enemy as each fragment acts like a virtual bullet. On the other hand, if an aircraft is to be destroyed or damaged, the fragments have to be bigger, each weighing at least about 5 g. The average velocity of the fragments may vary from as low as 1000 m/s in the case of HE shells to 4000 m/s in the case of large bombs. The shape, size, and velocity of fragments can be tailored by changing the explosive composition and the ratio of explosive-to-case weight. In a prefragmented ammunition, the size of the fragments to be formed is predecided by designing the outer shell accordingly.

What decides the effectiveness or lethality or range of the fragments? The effectiveness of fragmentation depends on three factors related to the explosive composition filled, viz.:

1. VOD of the explosive
2. Its density
3. Its power (i.e., its specific energy)

Immediately after detonation, the shock front exerts the entire detonation pressure on the shell. The nature of resulting fragments (their number, size, and velocity) depends on the detonation pressure, which again depends on factors (1) and (2) above. Just after the formation of fragments, they are propelled at high velocities by the high pressure gases produced, and factor (3) above plays the role here.

Kast has come up with the concept of "brisance values." To achieve high brisance, high VOD explosive compositions like RDX/TNT are used in fragmenting ammunitions. Different empirical methods have been reported in literature for comparing the brisance values of explosive, taking TNT as standard. The "Sand Test" determines the proportion of standard Ottawa sand crushed by 0.4 g of an explosive using standard sieves. The "Plate Dent Test" measures the dent produced on a standard steel plate by detonating a cylindrical column of explosive of known weight. "Upsetting Tests" measure the fragmentation effect of explosives by detonating a cylindrical column of the explosive kept over a lead or copper cylinder (keeping standardized buffer plates in between) and measuring the compression of the metallic cylinder after explosion.

4.4.2 Scabbing

If you want to damage an armor plate of a tank, can you do it without perforating it from its front side? Sound strange? Yes, it is possible if you exploit the "scabbing" effect of an explosive (Figure 4.7). It works like this:

The mechanism of scabbing is shown in Figure 4.8. When the explosive in close contact detonates, the plane shock front travels from left to right (front to rear of the plate) as a *compression wave*. The intensity or amplitude of the wave is proportional to its height. The moment the shock front reaches the rear end, it is "shocked" to see the total difference in the medium (steel to air) and, therefore, gets *reflected back as tension wave* (i.e., "negative shock wave," which instead of compressing the medium, does the opposite process of elongating it). At a particular place of the plate, as shown in the figure, the difference between tension and compression intensifies ($a_t - a_c$) and exceeds the fracture

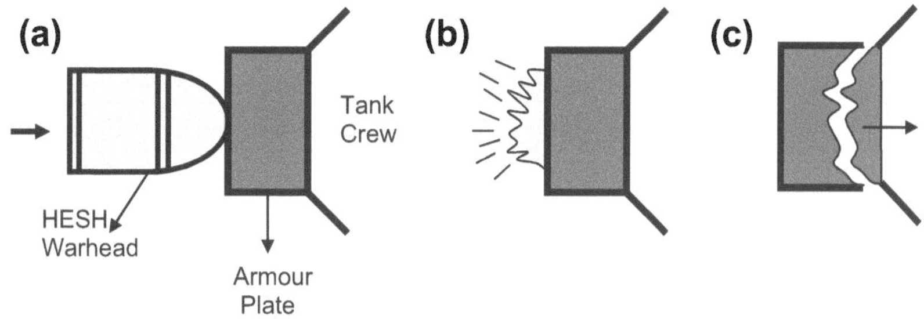

Figure 4.7
The Scabbing of an Armor Plate. (a) A High Explosive Squash Head (HESH) Ammunition Hits the Armor Plate. (b) The High Explosive inside the Warhead is "Spread" on the Plate, i.e., in Complete Contact with the Plate. It Goes Off. (c) A Large Chunk of the Armor Plate on Its ***Rear Side*** is Detached and Thrown Right into the Crew Cabin, Killing or Injuring the Crew and Damaging the Equipment.

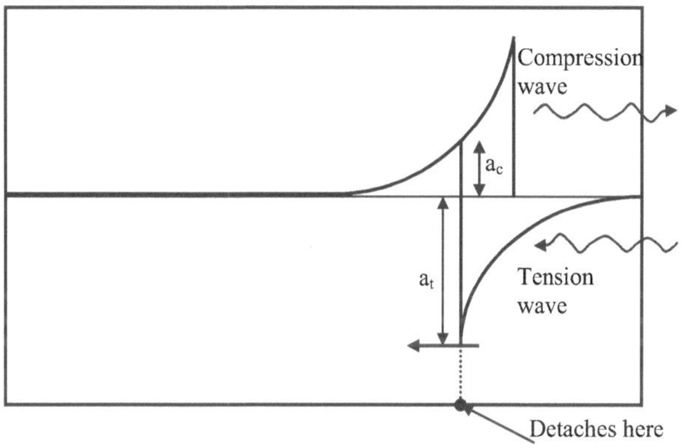

Figure 4.8
The Mechanism of Scabbing.

strength of the plate. At this point, a large chunk gets detached and is thrown from left to right, because the effective force here is not compression but tension. Depending on the quantity and power of the explosive and nature of the armour, the velocity of the scab can vary and can go up to 130 m/s, killing the tank crew instantly.

As scabbing is purely a shock phenomenon, the HE composition used in HESH ammunition must produce a high detonation pressure. Compositions based on RDX and HMX are mostly used for this purpose.

4.4.3 Shaped Charge Penetration

In 1888, C. E. Munroe discovered that when a conical cavity is created in a column of explosive charge, on detonation, it could penetrate a solid object. Further research in this

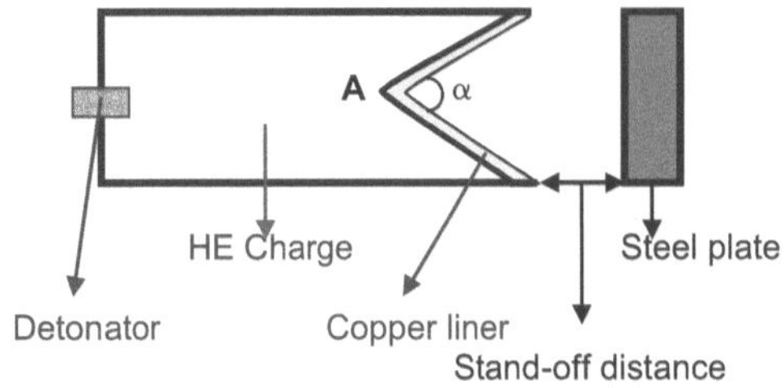

Figure 4.9
A Shaped Charge.

line established the ideal conditions for maximum penetration of hard targets like steel using such "shaped charges" or 'hollow charges.' The ideal conditions would be:

1. Introduce a concentric conical cavity at the end of a cylindrical column of an explosive charge.
2. Line the cavity with a malleable, metal-like copper
3. Keep a certain minimum "stand-off" distance between the base of the conical cavity and the steel plate (or target) as shown in Figure 4.9:

Once the detonation occurs, *the entire explosive energy is concentrated to collapse the metallic liner, which is transformed into a high velocity jet.* About 20% of the metal liner gets transformed into a jet, which has a velocity gradient from its tip (up to 9000 m/s) to tail (about 1000 m/s). A shaped charge, if properly designed, can penetrate steel plates eight times as thick as the diameter of the charge.

The mechanism of jet formation and penetration is interesting:

1. When the HE charge detonates, the shock wave passes over the liner, and the liner is accelerated at some small angle to the interface between the explosive and liner.
2. Since the liner velocity has been found to be directly proportional to the C/M (C: Charge weight of explosive over the liner, M = Weight of the metallic liner at a given point), it is observed that at the apex of the cone ("A" in Figure 4.9), the liner velocity is the highest. The liner velocity goes on decreasing as we go down the cone toward its base.
3. We, therefore, see that the jet comes out first from the apex and the rest of the liner is "squeezed out" following the "apex jet," but at lower and lower speeds. The remaining material, which is the bulk of the liner, forms a heavy slug that follows the jet at a much lower velocity, as shown below (Figure 4.10). In fact, after some point it becomes a discontinuous jet.

However, why do we need a stand-off distance? The stand-off distance (usually expressed as charge diameters) is essential for the proper development of a metallic jet until the tip

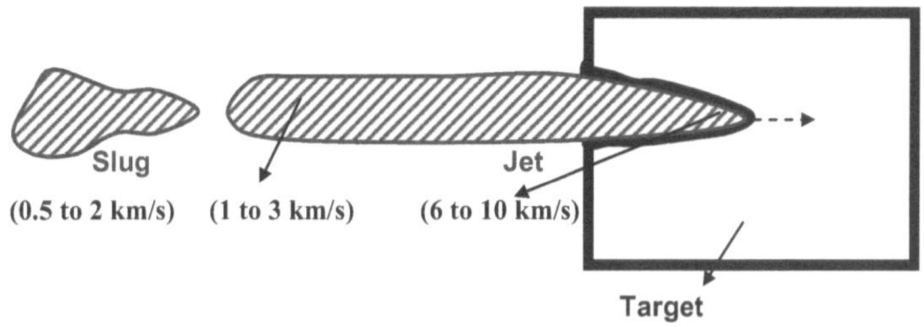

Figure 4.10
Jet and Slug Formation from a Shaped Charge.

achieves tremendous kinetic energy. The stand-off distance has to be optimal. If it is too short, we are not giving enough time and distance for the formation of high velocity jet. If it is too long, the jet breaks up and each particle hits further and further away from the center of the target. Maximum penetration has been found when the stand-off is about five times the cone diameter, with the cone angle around 42°. Optimum liner thickness is about 3% of the cone diameter for copper.

The penetration capacity of the liner is directly proportional to the kinetic energy of the jet and inversely proportional to the area of cross-section of penetration. As regards the explosive filled in a shaped charge, its VOD and density have a bearing on its performance. This is because of the fact that the main driving force behind the penetration is the detonation pressure (P_d), which is equal to $\rho D^2/4$, ρ, and D being the density and VOD of the explosive, respectively. Most of the shaped charges use RDX-based compositions. In exceptional cases, where still higher performance is needed, the much costlier β-HMX is used, as it has higher density and VOD than RDX.

The main applications of the "shaped charge effect" are their use in High Explosive Anti-Tank (HEAT) shells for the penetration of tank armor, in "cutting charges" used for the demolition of concrete and steel structures like bridges, and in a flexible, linear-shaped charge that consists of a flexible lead alloy with a Λ-shaped cross section that contains desensitized explosive composition for versatile applications.

4.4.4 Blast

Blast is defined as a phenomenon caused by rapid expansion of high temperature, high pressure gases as a result of an explosion. World War II operations proved that much greater damage can be done to installations by blast effect from bombs than the fragmentation effect. In fragmentation, the flying fragments damage only the points of impact, whereas a blast wave acts like a severe earthquake covering a circular area, damaging walls, throwing roofs and equipment, and also killing people if they are in the lethal overpressure region.

When an HE composition in a shell or a bomb undergoes detonation under confinement, the detonation pressure developed is in the order of a few 10^5 atm, with the temperature in the range of 3000–4000°C. The shell material gives way under such drastic conditions and about half of the explosive energy is used for this shattering and propelling of the fragments. The rest of the energy (due to high pressure and high temperature gases) is used for compressing the air around to form the shock wave with a steep plane front (See Chapter 3). The shock wave spreads spherically outward from the point of detonation and is followed by the gaseous products behind it. However, after a few milliseconds, the pressure falls to ambient pressure due to the inertia of moving gases (See Figure 4.11) at point A.

This is followed by a negative pressure (subatmospheric) phase from point A to point B. The major damage of the targets experiencing this blast wave (which consists of both the positive and negative pressure phases) is caused by (1) the peak pressure (represented by the height of the shock front OP), and (2) the area under the positive pressure phase (enclosed between OPA). This area is also known as the *impulse* of the blast wave, which is the (pressure X time) parameter, or the work done by the overpressure in "pushing" and damaging the object. In the second phase of negative pressure (area covered between A and B below X-axis), the gases go in the reverse direction, i.e., toward the point of detonation. At this time, the target experiences a "pulling" force rather than a push. This explains the fact that when a detonation occurs at a certain distance outside the window of

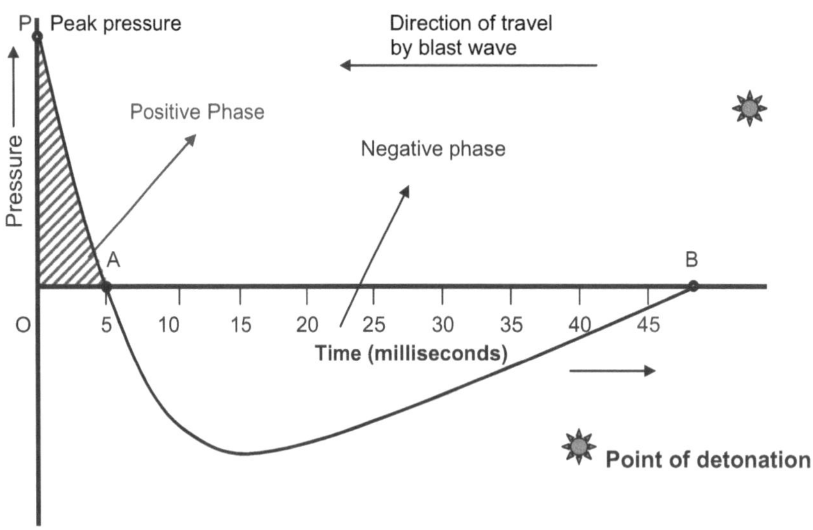

Figure 4.11
Two Phases of a Blast Wave.

a house, the window panes are shattered and the debris is found outside the house rather than inside.

Although both the peak pressure (OP) and the impulse (area POA) are destructive, their damage potential depends on the type of the target. If the target is a light and strong material (e.g., a window glass pane) a higher peak pressure is needed to break it. If it is a heavy but weak structure (e.g., a brick wall) a low peak pressure will do, but the impulse should be relatively higher. While an overpressure of 0.07–0.7 kg/cm^2 is enough to break a window pane, an overpressure of about 6 kg/cm^2 is needed to kill a man by fatally compressing his vital organs.

The blast effects depend on the medium and also the surroundings. While the blast wave attenuates quickly in an open space (its spherical area increasing in terms of $4\pi r^2$), it gets reinforced in confined spaces like a closed room due to its multiple reflection. While the blast effect is enhanced in a dense medium (e.g., underwater blast), it is severely reduced in a rarified atmosphere. For this reason, large quantities of HE composition are needed for efficient blast effect from the warheads used in anti-aircraft missiles at high altitudes.

4.4.4.1 Aluminized HE Compositions and Blast

Most of the HE compositions used for creating blast contain a certain percentage of fine aluminium powder, e.g., Torpex (41% RDX, 41% TNT, 18% Al), Tritonal (80% to 60% TNT, 20–40% Al), and Minol (40% NH_4NO_3, 40% TNT, 20% Al). *Aluminium plays a key role in extending the blast effect for a longer duration so that the impulse of the explosive increases.* The addition of aluminium powder produces a longer, flatter P-t curve in the positive pressure phase of the blast wave as shown in Figure 4.12.

It was found that aluminium does not participate in the initial detonation reaction, but it adds a large amount of heat to the entire system by subsequently reacting with the products of detonation, viz. water and carbon dioxide exothermically.

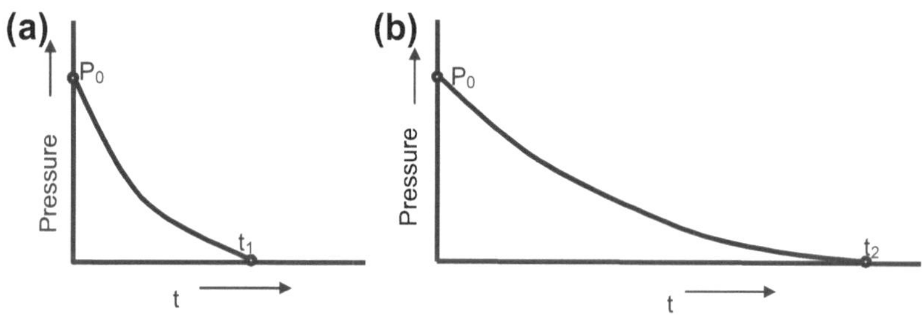

Figure 4.12
P-t Blast Profile of (a) Non-Aluminized Composition and (b) Aluminized Composition.

$$2Al_{(s)} + 3H_2O_{(v)} \rightarrow Al_2O_{3(s)} + 3H_2 + 207 \text{ k.cal}$$
$$2Al_{(S)} + 3CO_2 \rightarrow Al_2O_{3(S)} + 3CO + 177 \text{ k.cal}$$

The above reactions do not alter the total number of moles of gas in the system (Al and Al_2O_3 being solids), but at the same time, significantly increase the flame temperature due to the evolution of large quantities of the heat. This increased temperature increases the pressure of the gaseous products so that the positive pressure region gets further extended and the overall impulse (i.e., the area under the P-t curve) is enhanced.

However, increasing aluminium beyond a certain percentage is not preferable due to two reasons:

1. Firstly, excess aluminium may further react with CO to form carbon, thereby decreasing the total number of moles of gaseous products, resulting in the lowering of nRT values.

$$2Al_{(S)} + 3CO \rightarrow Al_2O_{3(S)} + 3C_{(S)}$$

2. Subsequent mixing of the unreacted or partly reacted gases with atmospheric oxygen may produce a delayed secondary explosion.
   The percentage of aluminium in an HE composition should, therefore, be optimized accordingly.

4.5 Industrial Explosives

4.5.1 Introduction

For more than 350 years, explosives have been employed to mine ores and minerals. World annual consumption of industrial explosives is at least 5×10^6 tons, a major part of which is ammonium nitrate—fuel oil (ANFO).

During the first 250 years of this period, only black powder was known and used, but fundamental changes occurred in the 1860s (the invention of dynamite and blasting cap by Alfred Nobel), 1950s (ANFO), and 1980s (emulsions). The search continues for less expensive products and safer techniques for production and field-use in mines, quarries, roads, tunnels, and dam construction. At the same time, the introduction of new products is restrained by the cost of existing investments and by safety and environmental regulations.

"Industrial Explosives" is too big a field to be discussed in detail in this section. The intention of the author is to sketch out only the salient concepts concerning them with respect to their chemistry vis-à-vis their application.

Gunpowder was, perhaps, the first civil explosive used. As far back as 1627, it was used for mining in Slovakia. Then came a less messy explosive in the form of dynamite

Figure 4.13
Alfred B. Nobel.

containing 75% NG, invented by Alfred Nobel (Figure 4.13). Other developments that followed included low-freezing NG explosives, Permitted Explosives (usable in gassy coal mines), detonators, and detonating cords. A new development took place in 1930s with the development of much safer, less sensitive and more cost-effective "blasting agents," mostly based on ammonium nitrate (AN). Blasting agents include "AN-Fuel Oil (ANFO)," non-cap sensitive "slurry explosives," and "emulsion explosives," which are of fairly recent origin.

The industrial explosives have played a great role in the development of the economy of many countries. The most common applications include mining, civil engineering, agriculture, petroleum engineering (seismic prospecting and perforation of oil wells), etc.

4.5.2 Requirements of Industrial Explosives

The sharp differences between the requirements of industrial explosives and that of military explosives have been mentioned in Chapter 1. The major requirements of industrial explosives can be summarized as under:

1. Safety of processing, handling, transport, and storage
2. Safety during their performance (for example, when used in gassy coal mines, they should not ignite marsh gas, resulting in a disaster).
3. Cost-effectiveness that calls for inexpensive starting materials.

4. Adequate strength
5. Tailorability of power
6. Good fume characteristics: no toxic gases should be evolved
7. Reasonable storage life
8. Good water resistance

The explosives chemist who has to formulate the composition for an industrial explosive for a specific use has a rather complex task ahead of him. He will have to choose a composition that will be the best compromise of the above requirements. He will have to use a variety of ingredients for this purpose. For example, he may have to use:

- an antacid like chalk to ensure better storage life of explosives based on NG, which undergoes acid-catalyzed decomposition during storage.
- Freezing point depressants like nitroglycol (NG freezes at 13°C and the solid NG is dangerously sensitive to impact and friction) to be added to NG-based explosives.
- Gelatinizers like NC to avoid exudation of NG.
- Flame temperature depressants like sodium chloride so that they can be safely used in gassy coal mines. Methane present in coal mines easily forms a dangerous explosive mixture with air. If the mining explosive results in a flame having a high temperature, longer duration, and higher length, it will ignite this explosive mixture, causing a disaster. NaCl ensures that a part of the explosive energy of the mining explosive is spent to dissociate it, thereby decreasing the temperature and duration of the flame and avoiding such a disaster.
- Fine powder of metals, particularly aluminium (but not in gassy coal mines where a methane–air mixture can be set off by the high temperature Al_2O_3 particles formed during the explosion) as a fuel-sensitizer.

Simultaneously, he has to ensure that all these ingredients are compatible with each other. He has to take care that the OB of the composition is almost zero. Highly positive OB and highly negative OB compositions will result in the evolution of toxic gases viz. oxides of nitrogen and carbon monoxide, respectively, and this will not be acceptable. Lastly, there should be no compromise in safety, storage stability, and the cost.

4.5.3 Industrial High Explosives

Today the industrial explosives can be broadly classified as "High Explosives" and "Blasting Agents." While the main ingredient in the former category is NG, the blasting agents mainly use AN in different types of formulations. NG-based explosives are quite powerful and more water-resistant than blasting agents, but the problem of their sensitivity, cost, and the limited extent of their tailorability have made them a thing of the past, paving the way for safer, cheaper, and more easily tailorable blasting agents.

The NG-based industrial explosives can be categorized as below:

1. *"**Straight dynamites**"* based on NG and kieselguhr. They are hardly used nowadays because of their high cost, sensitivity to shock and friction, and poor fuming characteristics (i.e., evolution of toxic gases during explosion).
2. *"**Ammonia dynamites**"* based on the introduction of less sensitive AN in dynamite composition. This results in lowered VOD and less objectionable shock sensitivity and fuming characteristics.
3. *"**Gelatine/Semigelatine explosives**"*: A small amount of NC is introduced to keep the explosive in gel/semigel form along with a variety of ingredients. A typical composition may include NG (base explosive), NC (gelatinizer cum explosive), AN (oxidizer cum explosive), saw dust (carbonaceous fuel), NaCl (flame temperature depressant), and chalk (antacid). These types of explosives can be used as "Permitted Explosives" (a term that is meant for explosives that can be safely used in gassy coal mines).

4.5.4 Blasting Agents

As the requirement of industrial explosives increased by leaps and bounds all over the world, there arose a necessity to look for NG-free explosives and substitute the NG-based explosives by AN-containing ones, as AN is the cheapest and safest source of readily deliverable oxygen for explosives. The concept of blasting agents originated from the United States. They are mainly AN-based, often free from high explosives like NG and TNT. They are cap-insensitive (a composition is said to be cap-sensitive if it would go off when initiated by a No. 8 detonator, which in terms of power, is equivalent to a 2 g mixture of mercury fulminate and potassium chlorate in the weight proportion of 80:20, respectively). Being insensitive, the safety regulations for their transport and storage are much less severe than other explosives in many countries. Some of the popular and widely used blasting agents are given below:

1. Ammonium Nitrate-Fuel Oil (ANFO)

In 1956, Prof. Cook showed that a 94/6 mixture of AN and a fuel oil based on saturated hydrocarbons represented an oxygen-balanced composition.

$$3NH_4NO_3 + \underset{\text{(Fuel oil)}}{CH_2} \rightarrow 7H_2O + CO_2 + 3N_2 + \text{heat}$$

This stoichiometric reaction yields maximum energy (1.025 k cal per gram of the explosive).

AN exists in five crystalline forms at different temperatures. At ambient temperatures, the transformation of one form to another changes the crystalline structure and unit cell volume. It is very hygroscopic and deliquesces at more than 60% relative humidity. This leads to caking of AN into virtually unmanageable "rocks of AN." This problem has been solved to a great extent by adding anticaking agents that are coated on AN crystals. Nowadays, AN is commercially produced as porous "prills" that are free-flowing and can absorb fuel oil uniformly. ANFO has replaced conventional explosives in open-pit mining. They can be mixed on-site simply by adding oil to a bag of AN prills. ANFOs, being cap-insensitive, are usually initiated with an HE booster such as pentolite (PETN/TNT:50/50). If needed, the sensitivity and energy of ANFO can be increased by the addition of explosives or powders of aluminium or ferrosilicon.

The principal disadvantage of an ANFO is that it cannot be used under wet conditions. This led to the development of "slurry explosives."

2. Slurry Explosives

Also referred as "watergel explosives" or simply "slurries," this class of explosives consists of:

1. A saturated aqueous solution of AN. This solution has a suspension of the following materials:
2. Undissolved nitrates like methyl ammonium nitrate.
3. Metallic fuels like aluminium
4. Organic fuels like glycols
5. "Sensitizers" like TNT or PETN in small amounts
6. A "thickener" like guar gum to impart cohesion or thickness to the entire composite mixture (like thickener added to soups).
7. Cross-linking agents like borax (which cross-links the −OH groups of guar gum, which is a polysaccharide. While guar gum swells due to the "unwinding" of its polymeric structure, cross-linking imparts rigidity to the watergel explosive).
8. Slurries are made sensitive to cap initiation by beating fine air bubbles into them. However, these have a tendency to coalesce on storage. Sometimes, "microballoons" (glass or polymeric bubbles with an average diameter of about 40 μm) are added to solve this problem. The air bubbles or microballoons increase the sensitivity of slurries to initiation due to the adiabatic compression of entrapped air in them that results in high temperatures and "hot spots."

Some of the advantages of slurries are:

a. They can be tailor-formulated to suit the energy and sensitivity requirements.
b. Very safe to process, handle and transport
c. Water-compatible (the swelling of guar gun provides an impervious sheath that offers water resistance to the explosive)
d. Good fuming characteristics
e. Wide choice of densities

Figure 4.14
An Avalanche being Triggered by an Explosion. *Photo Courtesy by Andrew Longstreth, Olympia Fire Dept., Bonney Lake, USA.*

f. Directly loadable by priming straight into boreholes.

3. Emulsion Explosives (or "emulsions")

Emulsions are fairly recent and have superior properties in comparison to ANFO and slurries. They are based on a "water-in-oil emulsion" system in which small droplets of saturated AN solution in water are dispersed in a mineral oil phase. This emulsion is stabilized by the use of surfactants. The advantages of emulsion explosives are:

1. As the fine droplets of oxidizer are in intimate contact with fuel, the explosive reaction is complete and it confers the advantages of low post-detonation toxic fumes, high VOD, and being highly waterproof.
2. High density
3. Higher oxygen balance
4. All these factors impart better blasting efficiency to emulsions. Emulsions can be made pumpable in large boreholes. They can be made in the form of cartridges also to replace the conventional NG-based explosives.

Apart from mining and quarrying, explosives have many other uses such as avalanche control, as shown in Figure 4.14.

Explosives in Avalanche Control

Loss of human life and property takes place every year due to snow avalanches. Artificial Triggering by firing the slopes is an economical and practical method of avalanche control and is used in various countries. Natural avalanches are triggered when the bonds that hold the snowpack together break from additional stresses created by factors such as rain, wind, rising temperatures, and the weight of new snow. It is difficult to predict exactly when or where an avalanche will occur. The most common way that avalanches are artificially triggered is through the use of explosives. The objective of avalanche control is to reduce or eliminate the hazard from potentially destructive avalanches. Avalanches may be initiated by detonating high explosives either in or above the snowpack. When such artificial triggers produce avalanches, impressions about snow stability can be ascertained, and options for avoiding the consequent hazards can be formulated. Explosives can be thrown by hand onto target zones or shot from mortar/artillery guns for more remote delivery of explosive charges.

4.6 Processing of the Compositions

Let us briefly see the principles based on which the explosives are processed. Firstly, explosives in pure form are hardly used for any application; rather, "**HE compositions**" containing one or more explosives and other ingredients are only used. The finished composition for military applications requires certain mechanical properties like machineability, and the pure explosives do not possess them. To get an explosive material of required mechanical property, thermal, and sensitivity characteristics, and also output, a composition of explosives containing certain ingredients is resorted to. Once this composition is decided, the technique of processing the composition is decided based on the physical properties of the ingredients. Three major methods of processing techniques are mentioned below:

4.6.1 Melt-Casting

TNT is the main explosive base in this technique. The major advantage of TNT is that its melting point is quite low (~81 °C), whereas its ignition temperature is much higher (240 °C). Other major explosives do not enjoy this advantage. For example, the melting point and ignition temperatures of RDX are 204 and 213 °C, respectively, and they are dangerously close. Unlike TNT, we cannot take the risk of melting RDX for making an RDX-based composition. Because of such a low melting point, the melting of TNT can be achieved by steam. High melting explosives like RDX or HMX can be incorporated in the molten TNT and cast in shells or bombs.

TNT has a very low OB (−74%). It is, therefore, mixed with explosives of higher OB like RDX (OB = −22%), which also adds to the VOD of the final composition, and with oxidizers like AN. Apart from the above, in some compositions, Al (to enhance the blast) and wax (to "phlegmatize," i.e., to desensitize) are also added. Some typical castables are given in Table 4.5.

Most of the castable explosives are machineable. Although the melt-casting process is simple and cheap, the final charges are sometimes prone to cracking, become sensitive, and cause settling of "heavy" ingredients during solidification, resulting in inhomogeneity in the composition.

Table 4.5: Composition and density of typical castable mixtures.

Explosive	Composition	Density (g/cm^3)
Amatol	TNT:60, AN:40	1.56
Composition B	TNT:39, RDX:60, wax: 1	1.713
Comp B2	40% TNT, 60% RDX	1.65
Torpex	TNT:40.5, Al:18, RDX:40.5, wax: 1	1.81
Octol	23.7% TNT, 76.3%HMX	1.809
Cyclotol	23% TNT, 77%RDX	1.743
Tritonal	80%TNT, 20%Al	1.72

4.6.2 Pressing

The crystalline forms of most of the explosives are such that they cannot be pressed as such. The pressed pellets may not have the desired cohesion or the crystals may be sensitive to friction or static electricity development during processing. A lubricant or phlegmatizer like wax is added to the composition before pressing. The pressing can be of different types: direct pressing (with or without vacuum), incremental pressing, or isostatic pressing. During the pressing, the pressure may be in the order of a few tons per square inch. The temperature and duration of pressing may also vary. It is possible in some cases to achieve a density of the final pellet almost close (about 99%) to the crystal density or sometimes called TMD.

4.6.3 Plastic Bonded Explosives (PBX)

PBXs are explosive compositions containing crystalline explosives like RDX or HMX to which polymeric binders have been added. The procedure of making a PBX is as follows:

1. The binder polymer is dissolved in a volatile solvent.
2. The explosive crystals/powder are added and mixed to form a slurry.
3. The solvent is evaporated, leaving a coating of the polymeric binder on the explosive particles.
4. These coated particles are die-pressed or isostatically pressed at high temperatures ($\sim$120°C) and pressures (1–20 kpsi) to get PBX pellets with densities very close to TMD.

Very high "solid loading" (percentage of solids like RDX in the overall composition) can be achieved, sometimes as high as about 97% in PBX compositions. A large variety of polymeric binders can be used like polyurethanes (Estane 5702-F1), polystyrene, fluoropolymers/copolymers (Viton A, Kel-F-800), nitroacetals/formals (BDNPA-F), etc. The major requirements of a binder for PBX are: thermal stability, low toxicity,

Table 4.6: Composition and density of a few plastic-bonded explosives.

Name of PBX	Empirical Formula	Composition	Density (g/cm³)
PBX-9010	$C_{3.42}H_6N_6O_6F_{0.6354}Cl_{0.212}$	90% RDX, 10% Kel-F	1.781
PBX-9011	$C_{4.406}H_{7.5768}N_6O_{6.049}$	92% RDX, 6% polystyrene, 2% DOP	1.69
PBX-9205	$C_{4.406}H_{7.5768}N_6O_{6.049}$	92% RDX, 6% polystyrene, 2% DOP	1.69
PBX-9501	$C_{4.575}H_{8.8678}N_{8.112}O_{8.39}$	95% HMX, 2.5% estane, 2.5% BDNPF	1.841
PBX-9404	$C_{4.42}H_{8.659}N_{8.075}O_{8.47}Cl_{00993}P_{0.033}$	94% HMX, 3% NC, 3% tris-β-chloroethyl phosphate	1.844
PBX-9407	$C_{3.32}H_{6.238}N_6O_6F_{0.2377}Cl_{0.158}$	94% RDX, 6% exon	1.61
PBX-9408	$C_{4.49}H_{8.76}N_{8.111}O_{8.44}Cl_{0.0795}P_{0.026}$	94% HMX, 3.6% DNPA, 2.4% CEF	1.842
PBX-9502	$C_{6.27}H_{6.085}N_6O_6F_{0.3662}Cl_{0.123}$	95% TATB, 5% kel-F	1.894

compatibility with explosive ingredients, ease of processing, safe and fast curing characteristics, and low glass transition temperature.

The main advantages of PBXs are their good mechanical properties, thermal stability, and safety in processing and handling. Some of the PBXs with their composition and density are listed in Table 4.6.

Suggested Reading

[1] T.L. Davis, The Chemistry of Powder and Explosives, Wiley, New York, 1956.
[2] M.A. Cook, The Science of High Explosives, Chapman & Hall, London, 1958.
[3] W. Taylor, Modern Explosives, The Royal Institute of Chemistry, London, 1959.
[4] T. Urbanski, Chemistry and Technology of Explosives, vols. 1–4, Pergamon Press, Oxford, New York, 1983.
[5] S. Fordham, High Explosives and Propellants, Pergamon Press, Oxford, New York, 1980.
[6] C.R. Newhouser, Introduction to Explosives, The National Bomb Data Center, Gaithersburg, USA, 1973.
[7] M.A. Cook, The Science of Industrial Explosives, IRECO Chemicals, Salt Lake City, UTAH, USA, 1974.
[8] F.A. Lyle, H. Carl, Industrial and Laboratory Nitrations, ACS Symposium Series No.22, Am. Chem. Soc, Washington, 1976.
[9] A. Bailey, S.G. Murray, Explosives, Propellants and Pyrotechnics, Pergamon Press, Oxford, New York, 1988.
[10] Blasters Handbook, Du Pont de Nemours, Wilmington, 1980.
[11] L.E. Murr (Ed.), Shock Waves for Industrial Applications, Noyes Publications, Park Ridge, New York, 1989.
[12] W.R. Tomlinson, Properties of Explosives of Military Interest, Picatinny Arsenal, Dover, N.J, 1971.
[13] C.E. Henry Bawn, G. Rotter (Eds.), Science of Explosives (Parts I & II), HMSO Publication, UK, 1956.
[14] Service Textbook of Explosives, Min. of Defence Publication, UK, 1972.
[15] Military Explosives: Issued by Departments of the Army and Airforce. Washington, DC, 1955.
[16] D.H. Liebenberg, et al. (Eds.), Structure and Property of Energetic Materials, Materials Research Society, Pennsylvania, USA, 1993.
[17] P.W. Cooper, Explosives Engineering, VCH Publishers, Inc., USA, 1996.
[18] C.E. Gregory, Explosives for Engineers, fourth ed., TransTech Publications, Germany, 1993.
[19] E.B. Barnett, Explosives, Van Norstrand Co., New York, 1919.

Questions

1. What decides the spontaneity of a reaction?
2. Coal gives more heat than TNT. But TNT detonates but not a piece of coal. Why?
3. How does the bond polarity of a functional group in an explosive molecule affect its sensitivity?
4. What are the two major parts of explosive energy?
5. Which major factors affect the VOD of an explosive?
6. What is meant by "Critical diameter" of an explosive? How is it explained?
7. The detonation temperature of PETN is 3400 K. Calculate its specific energy in J/g. (Ans: 984 J/g).
8. Why is lead styphnate added to lead azide in detonators?
9. Why are nitroaromatic explosives more stable than the nitric ester explosives?
10. What is the difference between an "igniferous train" and a "disruptive train"?
11. How can you increase the fragmenting power of an explosive?
12. What is the mechanism of scabbing action?
13. How does a conical liner collapse into a jet in an SC?
14. Why is the "stand-off" distance essential in an SC?
15. How does the addition of aluminium increase the blast effect?
16. What are the major requirements of an industrial explosive?
17. What are "Permitted Explosives"? Why is sodium chloride added in Permitted Explosives?
18. Why is the ratio of AN and fuel oil 94:6 by weight in ANFO explosives?
19. What is the role of (a) guar gum (b) microballoons in slurries?
20. Why are emulsions superior to slurries?
21. What are the three main types of processing explosive compositions?
22. What are the advantages of PBXs?
23. Give two examples for high melting point explosives.
24. What is the relationship between VOD, density and detonation temperature?
25. Define power index of an explosive.
26. Give some examples for "explosophores" and "auoxoploses."
27. Why is a high VOD explosive dangerous for coal mining application?
28. Who invented the dynamite and blasting cap?
29. Give some examples for castable explosives.

The Propulsive Facet of HEMs: I (Gun Propellants)

5.1 Introduction

Until the nineteenth century, gunpowder was widely used in most types of firearms. The invention of various smokeless powders led to the ultimate replacement of gunpowder as a propellant in rifles and guns. It was seen in the first chapter that a breakthrough was made by Alfred Nobel in the second half of nineteenth century by the invention of "Smokeless powder," by gelatinizing NC with NG. It was called "powder" as it was to replace the messy and inefficient "gunpowder" as a propelling charge. In fact, the propellants for small arms, mortars, and guns are in the form of "grains" of various shapes (solid cylinders, monotubular or multitubular or slotted-tubular cylinders, flakes, etc.) and sizes (as low as 1 mm in length and as high as a few centimeters) depending on the ammunition in which they are used. We will see shortly why we have to go in for such different shapes and sizes. As the propellant is meant to convert the chemical energy packed in it into mechanical/kinetic energy of the projectile, over the years, efforts were concentrated to develop propellants with higher and higher energy to propel projectiles of higher and higher masses to longer and longer ranges. At the same time, care was to be taken to control the flame temperature of the propellant and barrel pressure up to certain levels to avoid the erosion and bursting of the costly gun barrel, respectively. More than a century after the invention by Nobel, we have come a long way in the development of solid gun propellants for small arms, mortars, and guns of various calibers. The development of a gun propellant for a given ammunition for a given weapon is a joint exercise by the gun ballistician and the propellant chemist. While the former takes care of the physics of the drama inside the barrel during the propellant burning and projectile movement, the latter takes care of the chemistry, particularly the thermochemistry of the propellant ingredients.

5.2 Gun: the Heat Engine

Figure 5.1 gives a schematic representation of gun propulsion. "*W*" grams of the propellant inside the cartridge case (to which the projectile is crimped) would burn in a matter of a few milliseconds and the high pressure, high temperature gases would propel the projectile (a shot or a shell) weighing "*M*" grams, through the muzzle of the barrel.

Demystifying Explosives: Concepts in High Energy Materials. http://dx.doi.org/10.1016/B978-0-12-801576-6.00005-7

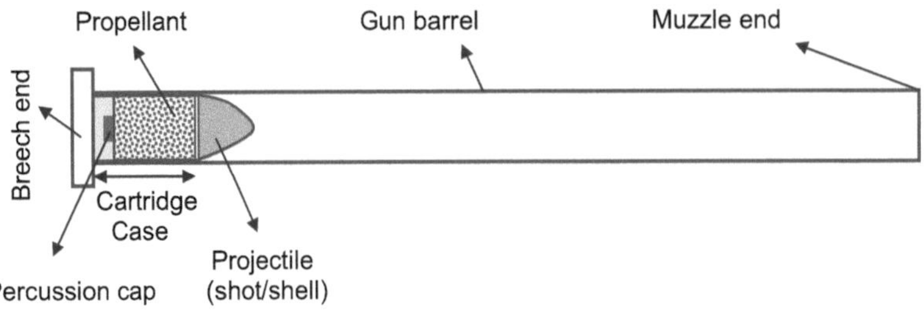

Figure 5.1
Schematic Diagram of Gun Propulsion.

The basic question is: how much of the chemical energy evolved (due to propellant burning) is converted into the kinetic energy of the projectile?

In thermodynamics, we call a system as a "heat engine" if it receives some heat from a "source," does some work out of it, and gives the balance to the "sink." If "Q" is the heat received, and W is the work done, then $(Q - W)$ is "wasted out" to the sink.

The efficiency of the heat engine is defined as the ratio of the useful work done (W) to the total quantity of heat it originally received from the source. When we apply this to a gun, we can realize that it behaves like a heat engine. The "source" is a propellant that gives total heat "Q" (which is the total heat produced by burning = Cal val × weight of the propellant) and the work is the movement of the projectile (or its kinetic energy). The wasted out energy $(Q - W)$ appears in terms of unutilized hot gases, heat transmitted to barrel walls, etc.

We can, therefore, write that:

$$\text{Efficiency of the gun} = \mathrm{e} = \frac{W}{Q} = \frac{1}{2}\cdot\frac{Mv^2}{Q}$$

where v is the velocity of the projectile.

As we know the value of Q and can measure the projectile velocity, v, we can calculate the value of "e" of a gun. The efficiency of a gun is found to be in the range of 30–45%. (There is no need to feel disappointed about this. This is a much better efficiency as compared to our automobiles, whose efficiency is never more than 20–25%).

A rough break-up of the distribution of the evolved energy is given below:

Mechanical energy	1.	For projectile motion = 42%
	2.	Friction = 3%
Thermal energy	1.	To hot gases = 29%
	2.	To barrel wall = 25%
Chemical energy	:	In unburnt propellant = 1%

(The above figures are reproduced from Ref. [11] given at the end. Although the efficiency figure of 42% quoted appears to be rather high, these figures give a rough idea about the propellant energy distribution.)

The Second Law of Thermodynamics states that heat can never be totally converted to work. This applies to the heat engine (gun) too. However, let us see which factors reduce the gun efficiency.

1. *Heat losses to barrel*: Proper design of the gun can minimize it but never eliminate it.
2. *Expansion ratio*: If V_1 and V_2 are the volumes of the product gases before and after expansion (i.e., the total volume of the barrel), respectively, assuming adiabatic conditions (although, strictly speaking it is not true, due to the heating up of walls), the efficiency of conversion of chemical energy to mechanical energy, "e" will be:

$$e = \left[1 - \left(\frac{V_1}{V_2}\right)^{\gamma-1}\right]$$

where γ is the ratio of specific heat of the gases evolved.

The more the gases expand, the better is the above conversion. If we need 100% efficiency, V_2 has to be infinity or we should have a barrel of infinite length.

3. *Pressure gradient*: There exists a pressure gradient in the barrel during the projectile movement. The pressure of the gases near the breech end (P_1) is far more than that is at the muzzle end (P_2). They are related as:

$$\frac{P_2}{P_1} = 1 - \frac{CZ}{2M}$$

where "C" is the propellant charge mass, Z is the fraction of the propellant burnt, and M is the mass of the projectile. The pressure gradient, which increases as the propellant burns and the projectile moves (causing the reduced efficiency as a higher pressure near the breech end), is not fully available to the projectile.

Worked Example 5.1

A gun has been designed for 35% efficiency. The ammunition of the gun contains 6.0 kg of a propellant of cal val 1050 cal g^{-1}. What muzzle velocity is expected of a projectile that weights 5.5 kg?

$$\text{The efficiency} = 35\% = 0.35 = \frac{1}{2}\cdot\frac{Mv^2}{Q}$$

[$M = 5.5$ kg, $Q = 1050$ cal g^{-1} $= (1050 \times 4.18 \times 1000)$ J $kg^{-1} \times (6$ kg$)$]

[Note: $J = kg\ m^2\ s^{-2}$, $J = 4.18\ cal\ g^{-1}$]

$$0.35 = \frac{5.5\,\text{kg} \times v^2}{2 \times (1050 \times 4.18 \times 1000)\,\text{J kg}^{-1} \times 6\,\text{kg}}$$

$$v^2 = \frac{(0.35 \times 2 \times 1050 \times 4.18 \times 1000 \times 6)}{5.5}\text{m}^2\text{s}^{-2} = 3351600\ \text{m}^2\text{s}^{-2}$$

Therefore, $v = 1831\ \text{ms}^{-1}$ is the expected muzzle velocity.

5.3 Unfolding Drama inside the Barrel

Figure 5.2 and the description below might help the reader to understand the sequence of events concerning gun propulsion. Gun barrel is the theater of this vivacious drama, which lasts for a few milliseconds. The first scene is the ignition of the propellant and the last scene is the exit of the projectile from the muzzle end. Figure 5.2 describes the change in the barrel pressure as well as projectile (or shot) velocity against shot travel.

1. The percussion cap at the base of the cartridge case is punctured by the striker pin. The forces of impact and friction ignite the pyrotechnic composition of the cap. This, in turn, ignites the propellant. It is assumed that all the grains of the propellant are simultaneously ignited, although it may not be exactly so.
2. The deflagration of the propellant results in the evolution of large amounts of high pressure gases within the cartridge. However, the projectile has certain inertia and also is crimped to the cartridge case. Only after the development of certain threshold pressure

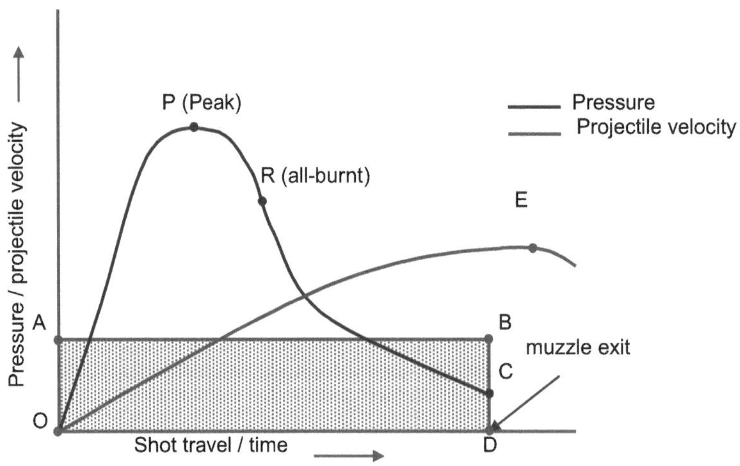

Figure 5.2
Pressure/Velocity—Time Profile inside a Gun Barrel.

(called Shot-start pressure), the projectile detaches itself from the cartridge case and starts moving along the barrel from point O.

3. We should understand that there are two types of pressure-time variation in the entire event. Firstly, it is the positive build-up of pressure in the barrel due to the continuous burning of propellant and evolution of gases, say, $+\left(\frac{dp}{dt}\right)_x$.

Secondly, as the shot moves, the gases have to expand, resulting in the reduction of pressure with the time, say $-\left(\frac{dp}{dt}\right)_y$. The main feature of the in-barrel drama is the competition between these two types of pressure variations. The net pressure-time, i.e., $\left(\frac{dp}{dt}\right)$ variation in the barrel, depends on which one of these is more dominating.

Initially, from point O to point P, there is a steep pressure rise due to the fact

$$\left(\frac{dp}{dt}\right)_x > \left(\frac{-dp}{dt}\right)_y$$

It is because, right from the word "go," the propellant starts burning promptly, whereas the projectile, due to its inertia, starts its acceleration process rather slowly.

4. At peak pressure, viz point P, they are equally competitive. At this stage, neither the propellant is fully consumed nor the projectile is out of the barrel.
5. From point P onwards, it is now the turn of the accelerating shot to outshadow the burning of the propellant so that:

$$\left(\frac{-dp}{dt}\right)_y > \left(\frac{dp}{dt}\right)_x$$

6. At point R, the entire propellant is burnt (called "all burnt" position) and the projectile has traveled only about one-third of its journey through the barrel.
7. At point C, the shot ultimately escapes from the muzzle but still gets further accelerated even beyond the muzzle up to point E (see the velocity curve) because of the muzzle pressure. Muzzle pressure is an important parameter in the design of the gun system because it gives that "extra kick" to the shot just when it is shunted out of the barrel! Similarly, the muzzle velocity is a vital parameter in gun ballistics.

The entire area under OPRCD represents the total work done by the gases to eject the projectile out. This area can be equated to the area of the rectangle OABD where OD is the time for the shot travel inside the barrel. OA is referred as the mean pressure of the barrel.

5.4 Energetics of Gun Propellant

It was said in Chapter 3 that the "mass burning rate," or sometimes called mass flow rate during burning denoted by "$\dot{m}$", is a very important parameter.

$$\dot{m} = rA\rho$$

Two parameters decide the value of $\dot{m}$ (apart from density ρ). Firstly, it is the ENERGETICS factor, i.e., the heat output (cal val) of the composition that decides value of "r". For example, if we take two identical strands of NG-based (high cal val) propellant and picrite-based (low cal val) propellant at a given pressure and temperature, the linear burning rate "r" of former will be much higher than that of the latter. The second factor is the CONFIGURATION. For a given composition (having a given value of "r"), if we make two grains of equal weight (same composition), one with larger surface area for burning (A) than the second, the former gets consumed much faster than the latter. If one measures the rate of rise of pressure of gases due to the burning of these two grains, (dp/dt) of the first will be higher than that of the second.

Let us take an interesting example. Picrite-based propellants (sometimes called "cool" propellants) are known to burn more slowly than NG-based propellants ("hot" propellants) as mentioned above. However, if we take two cartridge cases, the first containing 1 g of picrite propellant in the form of 1000 small cylindrical grains and the second containing 1 g of NG propellant in the form of 100 big cylindrical grains, on simultaneous ignition the former will burn out much earlier than the latter because of the larger surface area available for burning in case of picrite propellant, although it happens to be a cooler propellant.

We have seen in Chapter 2 that when a propellant burns, only a part of the evolved energy is diverted for the useful work of gas expansion PΔV, and the rest goes only to increase the internal energy of the gases (ΔE), i.e.,

$$Q = \Delta E + P\Delta V$$

In fact, PΔV should be substituted by PV because in ΔV ($\Delta V = V_{products} - V_{reactants}$) the volume of the gaseous products is far higher than that of the reactant, i.e., the solid propellants (about 1 g of the propellant occupying a volume of less than 1 cc evolves about 1000 cc of the gaseous products). Secondly, at such high pressures as we deal with in gun propulsion, the gases are no longer ideal in behavior and therefore, we have to correct the volume occupied by the gases with the co-volume factor "b" because of the significant value of the volume of molecules themselves at high pressures. The effective volume occupied by the gases will be (V − b) in place of V. During the deflagration of the propellant inside a gun barrel, these parameters are related as:

$P(V - b) = nRT_0$. (where T_0 is the flame temperature of the propellant)

If we substitute this in the above equation:

$$Q = \Delta E + nRT_0.$$

The following points need to be remembered with regard to the energetics of the gun propellants:

1. nRT_0 is the index of useful energy of a gun propellant. It shows how many joules of energy can be tapped from a burning propellant *exclusively* to propel a projectile. So, it has the units of J/g and is called the IMPETUS or FORCE CONSTANT of the propellant. For a given projectile weight and propellant charge mass, *the higher the value of* nRT_0, *the higher will be the muzzle velocity as well as the range of the projectile*. The maximum value of nRT_0 achieved in solid gun propellants today is of the order of 1300 J g^{-1}.
2. As T_0 is directly proportional to Q, *a higher cal val propellant achieves higher impetus.*
3. Similarly, if the propellant is based on a compound whose decomposition results in large values of "n" (number of moles of gaseous products per gram of the propellant) (low average mol.wt. of the product gases), the nRT_0 value goes up.
4. It is quite possible that propellant A has a lower cal val than that of B but has a higher force constant. The cal val. of NG and RDX are 1750 cal g^{-1} and 1360 cal g^{-1}, respectively. However, their impetus values are 1318 J g^{-1} and 1354 J g^{-1}, respectively. This is because 1 mole of NG evolves 7.25 moles of gaseous products, whereas 1 mole of RDX evolves 9 moles of the same. The increase in "n" value in case of RDX in comparison to NG has more than offset its lower cal val figure.
5. Beyond certain flame temperature, the gases start eroding the internal walls of the costly gun barrel. A limit for T_0 is, therefore, a must. Hence, the attempt of a propellant chemist is to formulate propellant compositions which have higher and higher values of "n" for an *optimized value of* T_0. Going by the above example, RDX-enriched propellants are preferred to the hot NG-based propellants.
6. The ratio of specific heats γ ($= C_p/C_v$) of the product gases influences the performance of the gun. We have seen in Section 5.2 that the efficiency of conversion of chemical energy to mechanical energy "e" is related as:

$$e = \left[1 - \left(\frac{V_1}{V_2}\right)^{\gamma-1}\right]$$

For a given expansion ratio, viz. (V_2/V_1) the efficiency increases as the value of γ increases.

For example, if we compare NG and RDX, the molar mean values of γ of their respective products of deflagration can be calculated (using standard values of γ available for CO, CO_2, H_2O, N_2, and O_2) as 1.3350 and 1.3773, respectively. If we substitute these values in the above equation, say, for an expansion ratio of 20 (i.e., $V_1/V_2 = 1/20$), it can be calculated that the values of "e" (efficiency) are 63.3 and 67.7% for NG and RDX, respectively. Thus, apart from the point of view of "nRT_0", RDX scores over NG in terms of "reduced inefficiency" due to expansion of gases inside the barrel.

The three parameters of energetics that matter for a gun propellant are, therefore.

1. T_0 = flame temperature (with an upper limit).
2. n = the no. of moles of the products per gram of the propellant.
3. γ = the ratio of specific heats of the product gases.

Worked Example 5.2

The impetus of picrite ($CH_4N_4O_2$, mol.wt. = 104.1) is 964 J g^{-1}. Calculate its adiabatic, isochoric flame temperature, T_0 (R = 8.314 J dg^{-1} $mole^{-1}$)

The deflagration of picrite is given as below:

$$\underset{(104.1\ g)}{CH_4N_4O_2} \rightarrow \underset{(5\ moles)}{CO + H_2O + H_2 + 2N_2}$$

104.1 g evolves 5 mol of gases.

Therefore, 1 g evolves 5/104.1 mol, i.e., n = 5/104.1 mol g^{-1}

i.e., n = 0.048 mole g^{-1}

Impetus (or force constant), F = 964 J g^{-1}

$F = nRT_0$, therefore, $T_0 = F/nR$.

$$T_0 = \frac{964\ J\ g^{-1}}{0.048\ mole\ g^{-1} \times 8.314\ J\ dg^{-1}\ mole^{-1}} = 2416\ K$$

(Note: This temperature is much lower than the T_0 values of many deflagrating explosives. That is why picrite-based propellants are also called "cool" propellants.)

5.5 Configuration of Propellant Grains

The pressure-time profile inside the gun chamber and the actual value of the peak pressure are very important. While the energetics of the propellant (nRT_0) matters a lot, the *rate of delivery of this energy* also matters equally. Imagine a high energy propellant in a cartridge case inside the barrel burning as slow as an incense stick for several minutes! Certainly, the projectile will never reach the end of the tunnel! The propellant grain configuration should, therefore, be optimally designed so that the required peak pressure is achieved within a matter of a few milliseconds to propel the shot promptly with the desired velocity.

If we imagine that the shot does not move, i.e., the volume available for propellant burning is a constant, for a given value of $\dot{m}$, the pressure of the gases will rise nearly linearly, i.e.,

$$\frac{dp}{dt} \propto \dot{m} \propto r\,A.\rho$$

However, let us remember that the shot is not stationary. It moves and the gases expand. Therefore, a constant value of $\dot{m}$ will not ensure a fast increasing value of (dp/dt); on the other hand, the (dp/dt) value might decrease if the rate of gas expansion is faster than the rate of its production. Therefore, the grain configuration is designed in such a way that the value of $\dot{m}$ increases with time. This is done by making the grain *progressively burning, i.e., with increasing surface area with time.* Mathematically:

$\dot{m} = rA\rho$ Assuming r and ρ are constants, $\frac{d\dot{m}}{dt} = (r\rho)\frac{dA}{dt}$

How do we achieve a progressive burning grain? Let us digress a little and see what is a "web" and what are the three modes of burning viz. regressive, neutral, and progressive burning.

Web is the minimum distance that can burn through as measured perpendicular to the burning surface.

A tubular grain is shown in Figure 5.3. When this grain is ignited, the burning proceeds from inside-to-outside (e.g., B → A) as well as from outside-to-inside (A → B). In this grain, the thickness AB, CD, etc. is the web of the grain as it represents the minimum distance that burns through. As the burning is two-sided in this case (viz. A to B, as well as B to A), the *effective web* will be AB/2 or CD/2. If the propellant burns at the rate of "r" mm s^{-1} and the web length (AB or CD) is "x" mm, the time taken to burn the entire grain will be $\frac{x/2}{r}$ s. Now let us see the three modes of burning:

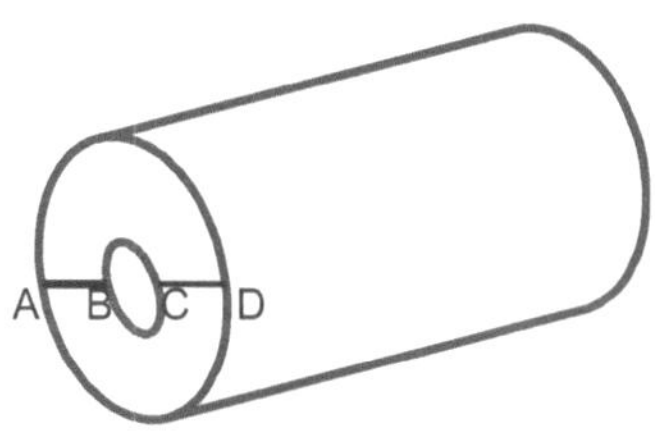

Figure 5.3
Web of a Grain.

5.5.1 Regressive Burning

If the surface area of the grain starts decreasing as the burning proceeds, it is called "regressive burning," e.g., a cord (i.e., solid cylinder). The P-t profile during such a burning is shown in Figure 5.4.

5.5.2 Neutral Burning

If the surface area of the grain remains same or nearly same during the burning, it is called "neutral burning," e.g., tubular grains. The P-t profile is shown below (Figure 5.5).

As the burning proceeds in a tubular grain, the increase in burning surface area due to inside-to-outside burning is compensated by its decrease due to outside-to-inside burning so that at any given time, the total available surface area for burning is same. Therefore, the value of does not change and the P-t profile is horizontal. Strictly speaking, although the changes in the peripheral areas of the cylinder (inside and outside) compensate each other, the areas at both the ends (shaded in the figure) decrease, thereby slightly decreasing the overall surface area of the propellant. This imparts a slight regressiveness making the burning "nearly neutral." This effect is reduced when the length-to-diameter ratio of the grain is increased. (In rocket propellants, as we will see later, the ends are "inhibited" by applying an inert polymeric coat so that they do not burn. This results in a perfectly neutral P-t profile.)

5.5.3 Progressive Burning

When we take a multiperforated grain like the heptatubular grain shown in Figure 5.6, it can be realized that the ignition starts simultaneously from all the seven holes as well as from the periphery. The rate of cumulative increase of surface area originating from seven holes far outshadows the rate of decrease of surface area due to burning from the periphery. As a result, the net surface area available for burning goes on increasing as the burning progresses as shown.

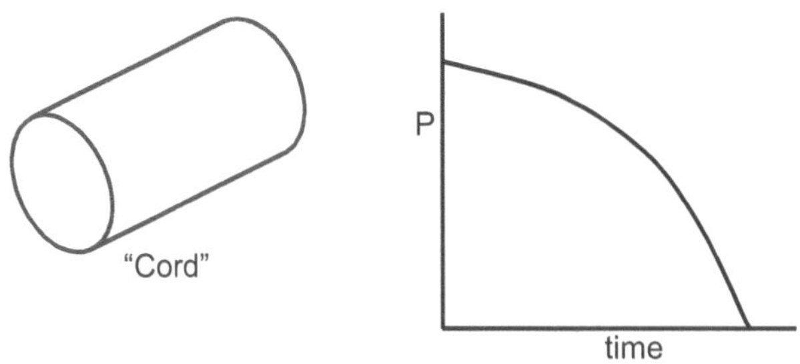

Figure 5.4
Regressive Burning of a Cord.

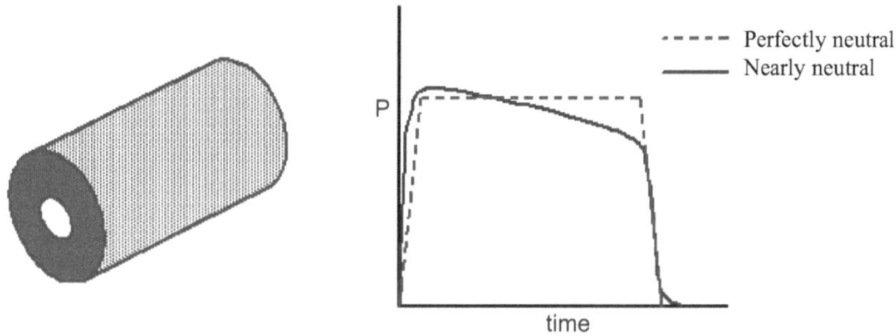

Figure 5.5
Neutral Burning of a Tubular Grain.

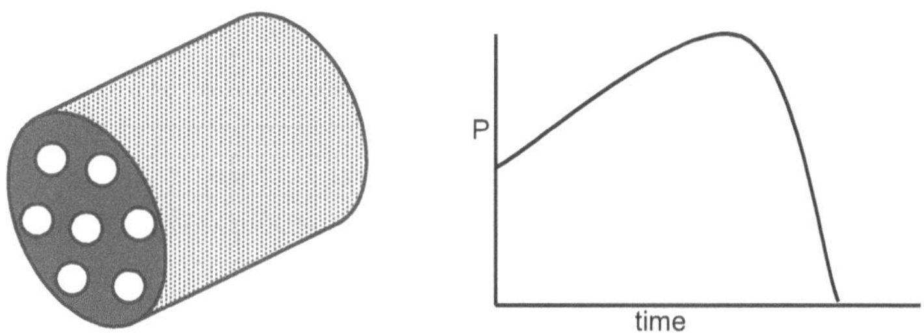

Figure 5.6
Progressive Burning of a Multitubular Grain.

Coming back to our problem of designing a progressively burning grain to achieve a high peak pressure (within the acceptable limits), we can see the reason why the multitubular geometry of the grains is common among gun propellants, especially for high performance guns such as tank guns.

One would find, particularly in small arms ammunitions, cord-type or spherical-shaped propellant grains (sometimes called "Ball Powder") that are obviously regressive burning. The reason is to be found in the fact that the barrels of these weapons are very short as compared to large caliber guns. There is just no time to "allow" the development of progressivity. The sense of urgency for the peak pressure development is much greater here. The cord or ball powder propellants have the maximum surface area right in the beginning to give the shock-kick to the projectile. However, care is taken to see that the pressure does not overshoot and burst the barrel by "moderating" the propellant grains by coating their surfaces with materials of negative cal val (e.g., phthalate esters that also act as plasticizers), and thereby keeping the burn rate under some check. It is like starting an automobile right on the fourth gear, but keeping a cautious pressure on the brake!

The foregoing two sections show that the two factors viz. energetics and grain configuration jointly decide how quickly the grain can burn, and this "quickness" is referred as the VIVACITY of the propellant.

5.6 Salient Aspects of Internal Ballistics of Guns

The term *"ballistics" means the study of the motion of a projectile.* "Internal ballistics" of gun refers to the branch of applied physics that deals with ballistic properties of propellants in relation to the motion of the projectile inside a gun barrel. Much of the theoretical work in this field started as far back as 1870s. In this section, it is not possible to give a detailed analysis of all the work done in this field, but a few salient points are mentioned.

1. The equations involved in the internal ballistics of gun establish the relation between the "gun parameters" (e.g., caliber of the gun, projectile mass, its velocity, its travel distance at any time "t", chamber volume, etc.) and "propellant parameters" (e.g., its cal val, force constant, web size and "form function" of the propellant grain, density of the grain, the pressure/temperature/ratio of specific heats/co-volume of the product gases, and the Equation of State for non-ideal behavior of the gases) by "Energy equivalence" equations, "Dynamic equations" (related to projectile movement), burning rate laws for propellants under ballistic conditions, and "form function."
2. *Burning rate law*: In 1885, Vielle established an important relation between the linear burning rate (r) of a propellant and the pressure (P) under which it burns as:

   $$r = \beta P^{\alpha}$$

   where α is the pressure exponent and β is the "burning rate coefficient" of the propellant. This law also applies to both the rocket propellants and gun propellants. The value of "α" may vary from 0.2 to 0.5 in the case of rocket propellants (actually, the symbol "n" will be used in the case of rocket propellants), and in case of gun propellants the value is in the range of 0.8–0.9. This higher exponential variation in the case of gun propellants is due to high pressures under which a gun propellant burns (about 4000–6000 kg cm^{-2}) as compared to rocket propellants, whereas P rarely exceeds 200 kg cm^{-2}.

   Higher pressures lead to:

   a. faster combustive chemical reactions.
   b. Faster heat transmission from the hot gaseous phase to the burning surface ("condensed phase").

A linear plot of "log r" against "log P" gives a straight line with "α" as slope and log β as its intercept on Y-axis.

While the value of α is almost a constant for gun propellants, *the value of β is quite characteristic of a propellant composition.* A higher value of β is undesirable for a given propellant composition as it may cause either uncontrolled burning or the problem of loadability. Let us consider two propellant compositions "1" and "2". Assuming their "α" values ~1, their burn rate equations are as given below:

$$r_1 = \beta_1 P, \quad r_2 = \beta_2 P$$
$$\beta_1 = \left(\frac{r_1}{P}\right) \quad \beta_2 = \left(\frac{r_2}{P}\right)$$

The units of "β" are expressed as cm s^{-1} $(MPa)^{-1}$ (where 1MPa ~ 10.1 kg/cm^2). If $\beta_1 >> \beta_2$, it means that propellant "1" has a much higher burning rate than propellant "2" at any given pressure. As a result, for a given grain configuration, its will be so high that the pressure generated within the barrel will exceed the safe specified limit. On the other hand, if we want to increase the web size of the propellant (thereby decreasing the total surface area per grain and hence), the available cartridge case volume may not accommodate the required charge weight of the propellant, i.e., the propellant becomes "unloadable." It should be noted that in a given volume of a cartridge case, the bigger the individual grains, the lower will be the quantity of the propellant that can be loaded in it.

Today, the value of β of many propellants lies in the range of 0.2–0.3 cm s^{-1} MPa^{-1}.

Worked Example 5.3

A gun propellant burning at a pressure of 500 MPa has the values of β and α as 0.25 cm s^{-1} MPa^{-1} and 0.92, respectively. Calculate the linear burning rate of the propellant at that pressure.

According to Vielle's Law:

$$r = \beta P^{\alpha} \quad (\beta = 0.25\ cms^{-1}MPa^{-1},\ \alpha = 0.92,\ P = 500\ MPa,\ r - ?)$$
$$\log r = \log \beta + \alpha \log P$$
$$= \log(0.25) + 0.92(\log 500)$$
$$= 1.8810$$
$$r = A\log\left(1.8810\right) = 76.03\ cm\ s^{-1}$$

3. Equations of State (EOS): The well-known EOS is PV = RT (for 1 mol of an ideal gas). No gas is ideal, and the non-ideal behavior increases at higher and higher pressures when:
   a. the volume of the molecules becomes significant when compared to the volume of the vessel which they occupy, necessitating a correction for their "co-volume" (denoted by "b" so that EOS becomes P(V−b) = RT).

b. due to further closeness between molecules, their intermolecular attractive forces increase, necessitating a positive correction for the real pressure they exert on the vessel by an amount = a/V^2.

The van der Waals equation thus takes the form:

$$\left(P+\frac{a}{V^2}\right)(V-b)=RT$$

("a" and "b" are called van der Waals constants)

It was argued by Abel and Noble that in the range of 2000–3000 K (the deflagration temperature range of gun propellants), the effects due to intermolecular forces can be neglected so that the above equation reduces to:

$P(V-b)=RT$ or $PV=k+bP$, (k = a constant), which is known as Noble–Abel equation. A plot of PV against P should give a straight line with a slope equal to "b". Mostly, the value of "b" is in the range of 1 ml g^{-1}.

The exclusion of intermolecular forces between molecules at high pressures does not really present the correct picture. Therefore, the Noble–Abel equation could not become the exact base for ballistic calculations. Numerous non-ideal gas equations were proposed, but most of them could not be applied for the gun ballistic conditions. The "truncated Virial equation," which takes into account the intermolecular potential (based on the method proposed by Lennard–Jones), was an improvement, although it too was not exact.

4. Computer Programs

Although many computer programs are available for performing thermodynamic computations of different reactions, only a few specific programs exist for the burning of gun propellants like TRAN 72 and BLAKE, which take into account of the non-ideality of gases more quantitatively to arrive at realistic solutions. They perform thermodynamic calculations including equilibrium concentration of gaseous products at constant, as well as varying pressures and temperatures under gun ballistic conditions. For more information, the readers may see the references given at the end of the chapter.

5. Closed Vessel (CV) Test

Gun firing using large amounts of gun propellants for the purpose of initial evaluation or for quality control during production is a costly affair. A CV apparatus is used for such purposes. The principle of a CV test is to fire a propellant of known loading density (i.e., a known mass of the propellant in a fixed space available inside the CV) using an "igniter" like gun powder and measuring the change of pressure (P) as well as (dP/dt) with respect to time. CV is a rudimentary laboratory tool that may not exactly replace a gun because (1) it does not exactly simulate the condition of gas expansion due to projectile movement as it happens in a gun barrel, and (2) the gases cool immediately after the firing in CV. However, CV firing can serve as a useful precursor before the actual gun firing.

The measurement of pressure versus time, (dP/dt) versus P, etc. is normally done for a propellant with reference to a "standard propellant" for the ballistic evaluation.

A typical (dP/dt) versus P curve of a gun propellant is shown in Figure 5.7. The standard (or reference) propellant and the candidate propellant are fired at the same conditions of loading density and temperature, and they are compared for two parameters viz.

a. *Relative Force (RF)*: which is a function of maximum pressure and that tells you about the total output of mechanical energy per gram of the propellant.
b. *Relative Vivacity (RV)*: which is a function of:

$$\left(\frac{dp}{dt}\right)_{max} \times \frac{1}{P_{max}}$$

(detailed equation are not given here) tells you how quickly or "vivaciously" the propellant burns, i.e., *the rate at which the mechanical energy is delivered.* As mentioned earlier, it is jointly decided by the energetics (nRT) factor of the composition and the configuration of the propellant grains.

One of the main factors defined in the interior ballistic calculations of guns is the "FORM FUNCTION." It defines the way in which the surface area of a particular grain shape changes during the course of burning. It is given by the following equation:

$$Z = (1 - f)\,(1 + \theta f),$$

where

Z = fraction of the grain burnt at time "t".

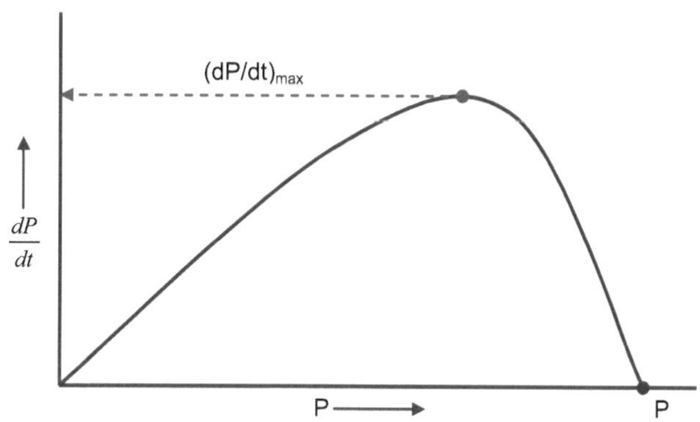

Figure 5.7
A Closed Vessel Firing Curve.

f = fraction remaining at the time of least thickness.

θ = form function.

The value of θ is zero for neutral burning geometries (e.g., Long tubes). It is positive and negative for grains that burn regressively (e.g., cords) and progressively (e.g., multitubular grains), respectively.

5.7 The Chemistry of Gun Propellant Formulations

The propellant chemist has rather a hard job on his/her hands. He/she is required to develop a propellant with suitable composition, shape, and size to meet the complex needs of an ammunition. It is not only the energy requirements he/she has to bother about. The propellant he/she develops should meet the following requirements in general:

1. *Energy delivery requirement*: In terms of cal val/nRT_0/loadability.
2. *Manufacturing characteristics*: In terms of cost and availability of raw materials/ hazards of manufacture/propellant viscosity and flowability/environmental considerations, reproducibility, etc.
3. *Storage requirements*: Effect of (low and high) temperature cycling on performance, mechanical properties, moisture absorption, exudation of plasticizer, etc.
4. *Compatibility requirements*: Compatibility with the process equipments and processing personnel (mainly toxicity) and compatibility among ingredients.
5. *Mechanical properties requirements*: To have good compression strength and percentage of compression at high (gun barrel) pressures. (If the grains crack under pressure before ignition, the extra surface area exposed will boost the barrel pressure to disastrous levels.) To withstand high acceleration forces and rough handling.
6. *Reliability of performance*: To ensure lot-to-lot reproducibility characteristics in terms of burning rate, RF, and RV.
7. *System requirements*: Smokeless and flashless exhaust gases, ignition and combustion stability, absence of pressure waves, absence of deflagration-to-detonation (DDT) characteristics, minimum sensitivity to heat, high velocity fragments, and other stimuli.

Very often, the achievement of all the above requirements at the same time may be quite difficult and the propellant composition is chosen as the best compromise of all these factors.

A gun propellant consists of the following main classes of ingredients:

1. "*Energetic binder*": To 'bind" all the ingredients into a cohesive grain and also impart energy (NC is the most commonly used binder).
2. *Plasticizers*: Energetic plasticizer like NG and fuel type plasticizers like phthalate esters.

3. *Stabilizers*: e.g., Carbamite, diphenylamine, etc.
4. *Coolants*: e.g., Dinitrotoluene.
5. *Flash suppressants*: e.g., potassium salts.

Today, we can classify gun propellants into four categories:

1. *Single base propellants*: Based mainly on NC. They also generally contain plasticizer, stabilizer, and flash suppressants (used in small arms and low caliber guns). Grain shape may be of cord or tubular, depending on ballistic requirements. Made by "solvent" extrusion method.
2. *Double base propellants*: (used mainly in low caliber guns and mortars).

Based on NC + NG gel matrix (more energetic than single base) + plasticizer + stabilizer + coolant—grain shape may be of tubular or multitubular or tiny spheres called "ball powder," flakes in case of propellant used for mortars. They are made by solvent as well as solventless extrusion methods.

3. *Triple base propellants*: Based on the (NC + NG + picrite) system containing similar additives as above—"cooler" (low flame temperature) and more "gassy" due to the presence of nitrogen-rich picrite (nitroguanidine)—used in large caliber guns. They are made by the solvent extrusion method.
4. *Low vulnerability ammunition (LOVA) propellants*: Propellant compositions excluding NC have been developed to impart insensitivity to accidental initiation of the propellant by high velocity projectile impact. They are based on inert polymeric binders like cellulose acetate in the matrix in which fine, desensitized RDX is dispersed to impart more impetus to the propellant. Some typical compositions and their performance parameters are given in Table 5.1.

(Abbreviation of the names of chemicals: DNT = Dinitrotoluene, DBP = Dibutyl phthalate, DPA = Diphenyl amine, DOP = Dioctyl phthalate, NC = Nitrocellulose, NG = Nitroglycerine.)

Except "ball powders," i.e., ball-shaped propellants, most of the gun propellants are made by the extrusion technique. The major steps involved in the processing of a typical single base propellant are given below:

1. *Dehydration*: Water-wet NC is dehydrated by mixing with alcohol and squeezing out in a press (Dry NC is highly sensitive to impact and heat and is, therefore, always stored with not less than 30% water in it).
2. *Incorporation*: NC (still containing a little water and alcohol) is mixed in a "sigma blade" mixer along with other ingredients. A calculated amount of ether and alcohol mixture is added at the time of this mixing to *partly "gelatinize" NC*. During the semi-gelatinization, the fibrous nature of the NC is partly destroyed. (The fibrous

Table 5.1: Composition and energetics of some typical gun propellants.

Parameter	Single Base		Double Base		Triple Base		Nitramine Base	
Composition (%)	NC (13.15%N)	90	NC (12.2%N)	49.5	NC (13.1%N)	20.8	NC (13.15%N)	30
	DNT	7.5	NG	47.0	NG	20.6	RDX	60
	DBP	1.5	Carbamite	3.5	Picrite	55.0	DNT	5
	DPA (+0.5 part K_2SO_4)	1.0			Carbamite	3.6	DOP	4
							Carbamite (+1 part K_2SO_4)	1
Cal val (cal g^{-1})	850		1175		880		1000	
Flame temp (K)	2850		3600		2793		3236	
Average molecular wt ($mole^{-1}$)	23.8		25.6		22.4		22.4	
Force constant (J g^{-1})	987		1168		1037		1190	
Linear burn rate coeff., β_1 (cm s^{-1} MPa^{-1})	0.10		0.25		0.13		0.15	

Note: The linear burn rate coefficient β is denoted as β_1 when it is assumed that $\alpha = 1$ in Vielle's equation.

NC burns too fast. In a finished propellant grain, this may result in the development of very high pressures and burst the gun barrel. If fully gelatinized, the burn rate will be too low to create the necessary peak pressure and P-t profile for imparting the required muzzle velocity to the shot. That is why we go in for semi-gelatinization.)

3. *Extrusion*: The mixed dough is extruded through a die-pin assembly to get long strands of required cross-section.
4. *Cutting*: The long strands are cut into grains of required length and dried well to bring down the solvent content (Volatile Matter%) as per specification.
5. *Graphiting*: The dried grains are given a fine coating of graphite with the following purposes:
   a. Graphiting ensures free flow of the grains and, therefore, better loadability in cartridge cases.
   b. Graphite, being a good conductor, helps in avoiding static electricity hazards that might accidentally initiate the propellant ignition.
   c. It helps in insulating the grain from ingress of moisture during storage.
6. *Sieving*: The grains are sieved to eliminate any odd-shaped or broken grains and fine powder.

7. *Blending*: Each batch is evaluated ballistically (e.g., by CV) and different batches are blended accordingly to realize the expected ballistics.

Three major factors that are to be taken care of during a propellant manufacture are:

a. *Quality Control*: Strict quality control needs to be exercised *right from raw material inspection to blending of finished batches.* For example, if NC has a lower "nitrogen content" (less percentage of nitrato groups in the chain), it will result in lower energy of the propellant. If its "ether-alcohol solubility" is more than specified, it might cause excessive gelatinization and reduction in burning rate of the finished propellant. Each and every process parameter is to be scrupulously respected to ensure the quality and reproducibility of performance of the finished propellant. (Sometimes, propellant making is described as an art. There is some truth in this statement, although each aspect or step of propellant processing has a scientific explanation. It is like giving the job of baking a cake to an experienced baker and a novice simultaneously. Although both of them know the finer details of the recipe and start with the same type of raw materials, the veteran baker comes out with a better cake! In the propellant processing, too, the experience plays a key role. For example, a veteran propellant processing technician knows by the look and texture of the dough whether the correct level of gelatinization has been reached or not.)
b. *Safety*: A baker can take a chance, but not a propellant technician! The latter deals with sensitive energetic materials and flammable solvents during the propellant processing. There can be no compromise with safety regulations during propellant manufacture like excellent housekeeping, flame-proof fittings, wearing of cotton clothes and conducting shoes (to dissipate any static electric charges), maintenance of the required relative humidity (min 60%), use of personnel protective equipments, strict adherence to the process schedule, etc. There have been a large number of instances when even a minor lapse in safety precautions resulted in disastrous accidents.
c. *Packing*: Proper packing of the propellant (both internal and external) as per the regulation not only ensures safety during transport and storage, but also ensures a long shelf-life of the propellant.

5.7.1 Role of Ingredients

The role played by some major ingredients used in gun propellants is described below:

1. Nitrocellulose (NC)

NC was synthesized more than a century back. Still, it rules the roost in many propellant compositions. That is because, its parent compound, viz. cellulose is a wonderful material. Cellulose is the natural polymer found in plants. It is a long polymeric carbohydrate chain interconnected by β-glucopyranose units. The molecular structures of cellulose and NC are shown below:

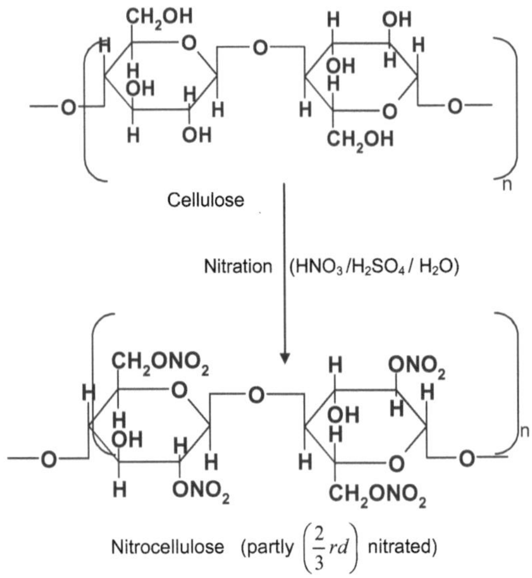

Each glycosydyl unit of the cellulose structure has three hydroxyl groups, viz. one primary OH (i.e., CH_2OH) group and two secondary OH (i.e., $-CHOH$) groups. Each unit of NC can be represented by the empirical formula ($C_6H_7O_2(OH)_3$). The square bracket shown in the figure covers two such units.

The cellulose polymer is a long chain with a large number of repeating units (n) and hence has a high molecular weight. The actual molecular weight of cellulose depends on the source and type of cotton linters or wood pulp from which it is prepared. Its molecular weight may vary from a few 100 thousands to a few millions. When cellulose is purified and nitrated using $HNO_3/H_2SO_4/H_2O$ mixture (called "nitrating mixture"), we get NC as shown above. Some of the interesting points in this connection are:

a. Depending on the end use of NC, the specification of properties of NC is varied. Some of the important properties of NC are (1) Nitrogen content, (2) Molecular weight (which determines the viscosity of NC when dissolved in a solvent like acetone or when gelatinized in a mixture of solvents like ether + alcohol), (3) average fiber length. The required properties depend on the source of cellulose and the nitrating conditions such as temperature, pressure, duration, and the actual composition of the nitrating mixture, as well as further processing of nitrated cellulose. For example, for the NC required for double base rocket propellants, we need NC with lower viscosity (and therefore molecular weight) as compared to NC for gun propellants. To achieve this, NC is pressure-boiled to breakdown the molecular chain of NC to a certain level.

b. It is very difficult to nitrate all the $-OH$ groups of cellulose to get fully nitrated NC (which theoretically corresponds to 14.14% N content).
c. *Nitrogen content (% N)*: By varying the nitrating mixture composition, the ratio of the nitrating mixture to cellulose, nitration temperature and nitration duration, NC with varying % N can be obtained. If x is the average number of nitrated groups (out of three in a unit) and y is the % N, we can show that and:

$$y = \frac{1400.8x}{162.14 + 45x} \qquad x = \frac{162.14y}{1400.8 - 45y}$$

Worked Example 5.4

Only 75% of the hydroxyl groups of cellulose could be nitrated during the manufacture of a batch of NC. Calculate the percentage nitrogen of NC obtained.

Every glycosydyl unit of cellulose contains three hydroxyl groups. The number of $-ONO_2$ groups in the final product (NC) corresponds to 75% of three OH groups, i.e., = 2.25 groups.

The above formula:

$$y = \frac{1400.8x}{162.14 + (45x)} = \frac{1400.8 \times 2.25}{162.14 + (45 \times 2.25)}$$
$$= 11.97\% \text{ is the nitrogen content}$$

An increase in % N (i.e., percentage of NO_3 groups) increases the energy (cal val) of NC. For example, the cal vals of NC samples with 12.60, 13.15, and 14.00% nitrogen contents are 3.91, 4.25, and 4.77 kJ g^{-1}, respectively. The percent N value of NC is, therefore, an important property, as that will be a decisive factor for the energetics and, to some extent, the mechanical properties of the propellants that are NC-based. The use of NC varies depending on its % N as shown below.

% N	Use
12.2–13.15	Propellants
11–12	Blasting gelatine
8–11.5	Commercial use (celluloids, lacquers, etc.)

d. *Viscosity*: Cellulose has a fibrous texture. After its nitration, NC still retains the fibrous texture although X-ray diffraction study shows a crystalline structure for NC of higher %N. The main characteristic of NC is its polymeric chain length, i.e., its molecular weight. During nitration of cellulose, the number of repeating units in the molecule gets reduced from 1000 to 3000 units (depending on the source and initial chemical treatment of cellulose) to somewhere between 400 and 700 units because of the molecular chain degradation owing to nitration conditions. The average molecular weight of NC plays an important role in propellant chemistry in terms of (a) *processibility*—for example, a high mol.wt. NC gives a highly viscous dough that cannot be extruded; and (b) *mechanical properties*: a lower molecular weight NC reduces the mechanical properties like tensile strength and compression strength of the finished propellant grain. It is, therefore, essential to have NC of optimum molecular weight.

The viscosity of standard solutions of NC (e.g., a given weight of NC dissolved in a solvent consisting a mixture of acetone and water in the ratio of 93:7 by volume, respectively) is indicative of its average molecular weight. Hence, the determination of viscosity of NC is an important quality-control aspect in a propellant manufacture. As mentioned above, the viscosity of NC can be brought down during its manufacture by "pressure-boiling" of its aqueous suspension in mild alkaline medium, and the process parameters need to be optimized and established to get NC of desired viscosity.

e. "*Blended NC*": During the manufacture of NC for small arms and gun propellants, there is a dual requirement. The NC sample to be used in propellant composition should have certain specified N% (let us say, N_X), and certain specified ether-alcohol solubility (let us say, S_X). While N_X ensures the correct energy level of the finished propellant, S_X ensures that NC will be gelatinized *to the required extent.* However, NC manufactured by nitration (called "straight cut NC") may not meet this dual need. Two different batches of straight cut NC (say having the values of N_1, S_1; and N_2, S_2, respectively) are blended in such a proportion that the blended NC meets the requirement of N_x and S_x.

2. Plasticizers/gelatinizers

The term "gelatinizer" is not to be confused with the term "plasticizer." Plasticizer facilitates the mobility of the molecules in relation to one another. Even inactive and inert compounds like Vaseline, which is a mixture of hydrocarbons, act as plasticizer. When they are added, say, during polymer processing, it increases the workability/flexibility/plasticity of the polymer apart from providing better low temperature properties like lower Glass Transition temperature to the final polymer product. *Gelatinizer*, on the other hand, interacts with the polymer by an electron donor/acceptor mechanism. Some compounds play both these roles as NG does with NC.

The fibrous texture of NC is mainly due to the interchain adhesion due to hydrogen bonding between the adjacent layers. In the case of NG, its molecules are small enough to penetrate through the interstitial space between NC layers and undo such interchain adhesion with the help of their own polar $-ONO_2$ groups. This helps in slidability of NC layers, thus effacing the fibrous texture of NC. What results is a gel matrix of NC/NG that becomes workable and safe, too. Thus, NG is a gelatinizer and also performs the function of a plasticizer.

Two major types of plasticizers are used in propellant manufacture viz. (1) energetic plasticizers (mainly NG) and (2) non-energetic/low energy plasticizers. Solvents like acetone and alcohol (containing polar groups of $C{=}O$ and $-OH$, respectively) are volatizable gelatinizers, i.e., they can be easily removed almost completely by the end of propellant processing. Phthalate esters (e.g., diethyl, diamyl phthalates) are nonvolatile plasticizers and are permanently present in the propellant composition. Phthalate esters also serve as fuels and have some stabilizing effect by absorbing any products of decomposition like oxides of nitrogen during the propellant storage.

3. Stabilizers

Being nitric esters, NC and NG have limited stability as the $RO{-}NO_2$ bond is susceptible to hydrolytic cleavage, resulting in the evolution of oxides of nitrogen over a period of time.

$$R{-}O{-}NO_2 \xrightarrow[\text{Higher storage temp.}]{H_2O(\text{moisture})} R{-}OH + HNO_3$$

$$2HNO_3 \rightarrow 2NO_2 + H_2O + (O)$$

Although the concentration of NO_2 evolved may be very small, it is sufficient to catalyze further decomposition of NC or NG, resulting in what is called the "autocatalysis" of the propellant decomposition. This is undesirable in terms of safety as well as the ballistic shelf-life (since loss of ONO_2 groups means loss of energy).

To arrest this possibility of the autocatalysis, some stabilizing compounds are added so that they can absorb in situ such oxides of nitrogen in their molecular structure and prevent the catalyzed decomposition of NC and NG. Some well-known examples of the stabilizers used in the propellant industry are given below:

a. Diphenyl amine (DPA)

DPA is a base and it absorbs the acidic oxides of nitrogen to form the nitro/nitroso derivative, thereby protecting NC from their attack. (DPA is used only in single base propellants. *It is not used in NG-based compositions* as it is too strong a base and initiates the base-catalyzed hydrolysis of NG.)

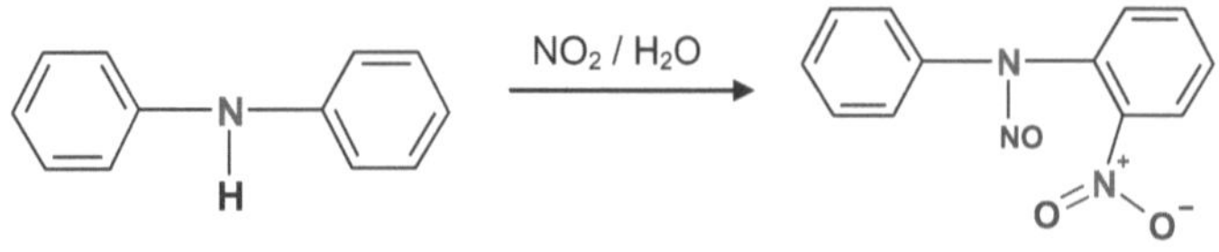

DPA N-nitroso, 2 nitro DPA

b. 2-nitro diphenyl amine (2NDPA)

As shown below, the nitro group of 2NDPA, due to its electron withdrawing tendency, reduces the basicity of DPA. NG-*containing compositions, therefore, use* 2NDPA *as stabilizer.*

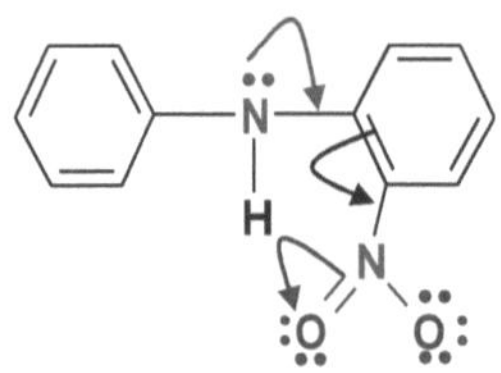

c. Sym-diethyl diphenyl urea (also called carbamite or ethyl centralite)

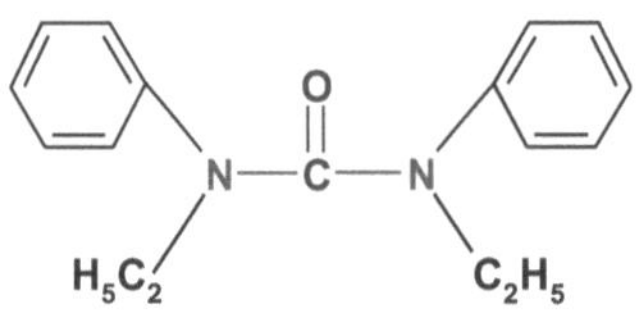

It is an excellent stabilizer, which readily absorbs any evolved oxides of nitrogen. It also acts as a plasticizer and a moderant.

4. Antacids (e.g., chalk)

NC-containing propellants are likely to have "acidity problems" originating from the manufacture of NC, wherein strong nitrating mixture is used. Chalk ($CaCO_3$) in small quantities is added to neutralize this acidity and prevent any acid-catalyzed decomposition of NC and NG in propellant composition during storage.

5. Coolants

Compounds with low cal val are added to propellant to bring down the flame temperature of propellants in certain compositions. These compounds endothermically decompose and thereby, reduce the overall heat output during propellant deflagration. Dinitrotoluene (DNT) and phthalate esters act as coolants.

6. Flash suppressants

NC, the major ingredient in most of the gun propellants, has a negative OB. All other ingredients, with the exception of NG, have a still higher negative OB, with the result that the propellant composition, as a whole, always has a negative OB. As a result, when the propellant deflagrates within the barrel, the product gases that come out are severely underoxidized and abound in CO and to a fair extent, H_2. The deflagration of NC (with 12.75% N) can be written as:

$$2C_{12}H_{15}O_{20}N_5 \rightarrow 6CO_2 + 18CO + 10H_2O + 5H_2 + 5N_2 + \text{Heat}$$

($C_{12}H_{15}O_{20}N_5$ refers to the empirical formula of one repeating unit, i.e., with two glycosydyl units in NC molecule, where five out of six hydroxyl groups of its cellulose precursor have been replaced by $-ONO_2$ groups.)

When large quantities of hot and oxygen-hungry gases of CO and H_2 rush out of the muzzle, they are greeted by the atmospheric oxygen and immediately get oxidized to CO_2 and H_2O, respectively. As these reactions are highly exothermic, the heat of their combustion appears as a big flash. (The Lower Explosive Limits of CO and H_2 in air are 12.5 and 4%, respectively).

Such a big muzzle flash in a battle scenario is undesirable as it reveals the position of the gunner to the enemy, particularly in the nights. To suppress the muzzle flash, salts of potassium such as K_2SO_4, KNO_3, and K_3AlF_6 are added in the propellant composition. Studies have revealed that at the high deflagration temperatures of the propellant, these salts decompose to form the free radicals of the potassium metal that, being highly reactive, immediately combine with oxygen and thereby inhibit the chain reactions that are responsible for the oxidation of CO and H_2. At the muzzle condition, the preference of atmospheric oxygen goes to higher reactive potassium free radicals rather than CO and H_2. One disadvantage of these inorganic salts is that, although they suppress the flash, they cause some amount of smoke.

It is relevant to make a mention about the role of nitroguanidine (picrite) in triple base propellants that, for instance, are used in the large caliber gun ammunitions. Picrite has two advantages viz. (1) it is very rich in nitrogen (53.8%) with the following structure:

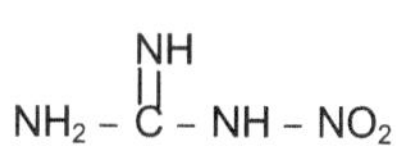

Large amounts of nitrogen in the product gases dilute the CO and H_2 and reduce the chances of their oxidation and generation of flash (2) It is a "cool" ingredient (cal val $= 769$ cal g^{-1}) and hence, the flame temperature of the propellant is low. This significantly reduces the barrel erosion and enhances the barrel life.

(A question might arise in the light of above description: Why then, do we not go in for a propellant composition that has zero or positive OB so that we avert the formation of CO and H_2? Firstly, such a composition will generate more of CO_2 and H_2O and increase the average molecular weight of gases, or decrease the value of "n", thereby decreasing the impetus of the propellant. Secondly, the complete oxidation being much more exothermic than in the case of production of underoxidized CO and H_2, means that the flame temperature of the products will rise to unacceptable levels and cause severe barrel erosion). (3) When the OB of the propellant moves closer to zero, there is a tendency for the propellant to undergo DDT within the barrel, and this would be disastrous.

7. Surface moderants

It was mentioned in Section 5.5 (under "regressive burning") that certain propellant grains, particularly those of regressive burning type, should be surface-coated with substances that decelerate the initial burning rate of the propellant. Substances such as DNT, phthalate esters, carbamite, etc. are useful for this role. They should have either a very low or negative cal val and should be nonvolatile. They are dissolved in ethanol and the solution is sprayed on the propellant grains in a sweetie-pan. Subsequently, the solvent is removed by heating, leaving a thin layer of the moderant on the surface of the propellant grains.

8. Wear reducers

A gun barrel is a costly material. It is a product of precision engineering and made of costly alloy. It has to withstand high pressures and high temperature gases, from round to round. Beyond certain increase in caliber, the barrel has to be abandoned as the sealing of product gases, for the generation of required pressure will cease to operate. Efforts are, therefore, directed to incorporate certain ingredients either in the propellant composition or in a "wear-reducing liner" that is inserted in the cartridge case before loading the propellant in it. Some of the anti-wear additives used are TiO_2 and talc, which are naturally occurring magnesium silicate. These compounds are "waxed" to the surface of the anti-wear liners and when the propellant deflagrates, the wax melts and a fine layer of TiO_2 or talc gets deposited on the inner walls of the barrel. TiO_2 and talc, being excellent heat insulators, are contained by a layer that protects the barrel walls from hot gases to a great extent. This layer gets removed when the next round of ammunition is fired, but then, a fresh layer is formed. This cyclic process of layer formation, thermal insulation from hot gases, layer removal, and layer reformation goes on. Eventually, it increases the barrel life.

9. Decoppering agents

Many of the gun barrels have "rifled bore," i.e., they have grooves made inside, commencing from a certain distance from the muzzle end. These grooves impart a high spin to the moving projectile because a spin-stabilized projectile has a better aerodynamic stability during its travel from the muzzle to the target. The fast spinning, high-speed projectile causes great frictional force between the barrel and the driving band of the projectile. This results in the deposit of fine copper particles from the driving band into the grooves, which is undesirable both in terms of safety and ballistics.

To solve this problem, compounds of lead and tin are added in small amounts in the propellant composition. During the propellant deflagration at high temperatures, these compounds decompose and form a low melting, high density alloy of lead and tin that flushes out the fine copper deposits from the grooves.

Suggested Reading

[1] S. Fordham, High Explosives and Propellants, Pergamon Press, Oxford, New York, 1980.
[2] K. Fabel, Nitrocellulose, Enka, Stuttgart, 1950.
[3] F.D. Miles, Cellulose Nitrate, Oliver & Boyd, London, 1955.
[4] J. Quinchon, J. Tranchant, Nitrocelluloses, the Materials and Their Applications in Propellants, Explosives and Other Industries, Ellis Howard Ltd, Chichester, UK, 1989.
[5] R. James, Propellants and Explosives, Noyes Data Corporation, Parkridge, New Jersey, 1974.
[6] R. Krier, et al. (Eds.), Interior Ballistics of Guns, Progress in Astronautics and Aeronautics, vol. 66, AIAA, New York, 1979.
[7] C.L. Farrar, D.W. Leeming, Military Ballistics, a Basic Manual, Brassey's Publishers Ltd, Oxford, 1983.
[8] Internal Ballistics, HMSO Publication, UK, 1951.
[9] L. Stiefel (Ed.), Gun Propulsion Technology, Progress in Astronautics and Aeronautics, vol. 109, AIAA, New York, 1988.
[10] Service Textbook of Explosives, Ministry of Defence Publication, UK, 1972.
[11] E.D. Lowry, Interior Ballistics, Doubleday & Co., Inc, New York, 1968.
[12] J. Corner, Theory of Interior Ballistics of Guns, John Wiley & Sons Inc, 1950.
[13] W.C. Nelson (Ed.), Selected Topics on Ballistics, Pergamon Press, London, New York, 1959.

Questions

1. What is the order of efficiency of a gun? Which factors affect the efficiency?
2. In an anti-tank gun ammunition, 5.1 kg of a double base propellant whose cal val is 1100 cal g^{-1} is used. If the projectile of this ammunition weighs 5.2 kg and achieves a muzzle velocity of 1440 ms^{-1}, calculate the efficiency of the gun. (Ans: 23.0%).
3. Why is muzzle pressure an important parameter?
4. What is meant by "Impetus" or "Force Constant" of a propellant?
5. The average molecular weight of gases produced during the deflagration of a propellant is 21. If the adiabatic isochoric flame temperature reached during the deflagration is 3000 K, calculate the impetus of the propellant. (Ans: 1188 J g^{-1}).
6. How do the shape and size of a propellant grain influence the rate of pressure rise inside a gun barrel?
7. The propellant grain configuration in some cases is meant for progressive burning, whereas in some others for regressive burning. Why?

8. What is Vielle's Law? Why should one be concerned about the value of the burning rate coefficient of a gun propellant?
9. A gun propellant, burning at a pressure of 400 MPa, has the values of β and α as 0.2 cm s^{-1} MPa^{-1} and 0.90, respectively. What is its linear burning rate at that pressure? (Ans: 43.95 cm s^{-1}).
10. What is the purpose of a closed vessel? What do you understand by relative force and relative vivacity?
11. What are the major requirements of a propellant?
12. Distinguish between single base, double base, triple base, and nitramine base propellants.
13. Why are certain propellant grains graphited?
14. Why is it said that the propellant making is an art?
15. Nitrogen content of a sample of NC is 13.00%. Calculate what percentage of hydroxyl groups of its precursor (cellulose) has been nitrated. (Ans: 86.13%).
16. Distinguish between a plasticizer and a gelatinizer. Why do we "semi-gelatinize" NC while processing propellants for small arms?
17. Why do we blend NC batches before the processing of gun propellants is commenced?
18. Write the possible chemical equation to explain the mechanism of how carbamite addition stabilizes a propellant composition.
19. In double base propellants, DPA cannot be used as a stabilizer. Why?
20. What is the mechanism of production of muzzle flashes and also their suppression using inorganic salts in the propellant composition?
21. What are the roles of (a) Surface moderants, (b) Wear reducers, and (c) Decoppering agents in a propellant composition?

CHAPTER 6

The Propulsive Facet of High Energy Materials—II (Rocket Propellants)

6.1. Introduction to Rocketry

The Chinese are credited to have invented rockets several centuries back. Gunpowder-filled paper tubes sealed at one end with a wick on the other were known to propel themselves on ignition, soaring toward the sky against gravity. What started as a part of firework display in the early stages found its application in modern missiles and space missions during the last century. Today, the load carried by a rocket (commonly known as the "payload") can be either a warhead—conventional or nuclear—or a satellite that needs to be "injected" into a particular orbit of the Earth for communication purposes. Thus, rockets have become part and parcel of modern life for various applications such as entertainment or war or space research. Although long-range missiles with nuclear warheads threaten the very existence of mankind today, global space research programs hold great promise for advancement in various fields such as communication, weather prediction, and tapping the resources from Earth. When the famous U.S. astronaut Neil Armstrong created history by becoming the first human to set foot on the lunar soil on July 21, 1969 ("A small step for me but a giant leap for mankind"), his ecstasy and excitement were shared by several millions on Earth. Thus, the field of rocketry has become an inalienable part of today's science and technology. The aim of this chapter is just to introduce the basic principles of rocket propulsion and the role played by high-energy materials (HEMs) in the form of rocket propellants toward propulsion performance.

6.2 Basic Principles of Rocket Propulsion

A rocket motor basically consists of two parts: a propellant combustion chamber and a nozzle (see Figure 6.1). The chamber is a metallic tube sealed at one end and the rocket propellant (in the case of solid rocket propellants) is loaded through the open end. The propellant grain may be of varying shapes and sizes depending on the type of performance expected from the rocket. For example, it can be a solid cylinder or a tubular propellant grain, as shown in Figure 6.1. The annular space in the tubular propellant grain is called the "port." The loaded rocket chamber is then screwed onto a nozzle, which in most of cases is a convergent-divergent (CD) nozzle, as shown in Figure 6.1. An igniter placed in

Demystifying Explosives: Concepts in High Energy Materials. http://dx.doi.org/10.1016/B978-0-12-801576-6.00006-9

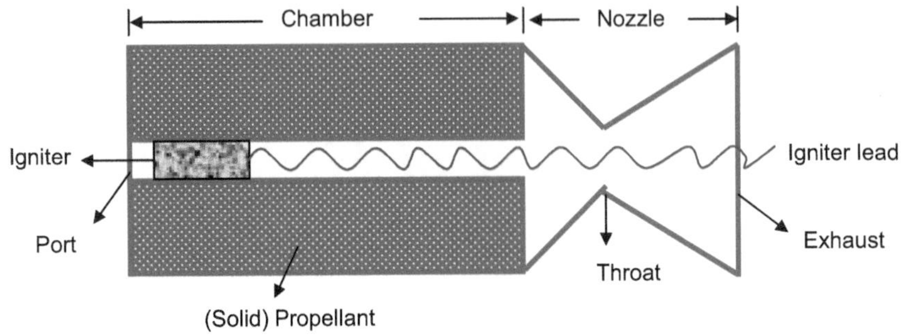

Figure 6.1
A Rocket Motor.

the port of the motor initiates the ignition of the entire propellant surface. This results in the production of high-temperature and high-pressure gaseous products, which get accelerated to very high velocities with the help of the nozzle. There is a tremendous increase in the velocity of the gaseous products when they expand from the "throat" portion of the nozzle to its exit. It is a matter of common knowledge that the exiting gases "kick back" the rocket as per Newton's Third Law of Motion, thereby resulting in propulsion.

The total thrust (F) with which a rocket is propelled has two components (Figure 6.2).

The first component of F is due to the thrust (F_1) created due to the imbalance of chamber pressure (P_c) and exhaust gas pressure (P_e) acting on the throat, the area of which is A_t. Therefore, it can be written,

$$F_1 = (P_c - P_e)A_t \tag{6.1}$$

(Note: Because the high-pressure gases are expanding after passing through the throat, P_c is always much greater than P_e; therefore, F_1 always has a positive value.)

Figure 6.2
Components of Rocket Thrust.

The second component of F is due to the thrust (F_2) created due to the imbalance of exhaust gas pressure (P_e) and the ambient pressure outside of the rocket (P_a) acting on the exhaust, the area of which is A_e. Therefore, we can write

$$F_2 = (P_e - P_a)A_e \tag{6.2}$$

(Note: P_e is often greater than P_a (called the "underexpanded nozzle") so that F_2 also has a positive value. At times, the rocket is designed in such a way that $P_e = P_a$, resulting in $F_2 = 0$ (called the "optimum expanded nozzle").

There is also a possibility that $P_e < P_a$, as it happens when the nozzle becomes longer (resulting in an "overexpanded nozzle"), thereby resulting in negative values of F_2. P_a is not just the atmospheric pressure but the ambient outside pressure. For instance, when the rocket sails through vacuum in the interplanetary space, P_a is almost equal to zero and F_2 assumes the maximum value. The net propulsive force (F) a rocket experiences is the sum of F_1 and F_2; that is,

$$F = (P_c - P_e)A_t + (P_e - P_a)A_e \tag{6.3}$$

6.2.1 Types of Rocket Engines

A rocket is basically an energy conversion system converting the stored chemical energy in a propellant to the kinetic energy of the exhaust gases through nozzle expansion. One of the methods of classification of rocket engines is based on the physical status of the propellant. They are basically classified as

1. Solid propellant rockets,
2. Liquid propellant rockets, and
3. Hybrid propellant rockets.

The simplest rocket engine in the design and working point of view is a solid propellant engine, and most of this chapter describes solid propellant-based rockets. The solid propellant mostly consists of a mixture of an inorganic oxidizer (most commonly ammonium perchlorate (AP)) and a metallic fuel (e.g., aluminum) embedded in a matrix of polymer, which performs the dual functions of a binder (giving structural integrity of propellant grain) and fuel. Such a propellant is called a "composite propellant." For certain military applications, double-base rocket propellants (DBRPs) based on nitrocellulose (NC) and nitroglycerine (NG) are still being used.

As the name implies, the liquid propellant engine consists of a propellant that is a liquid. Again, there are two types of liquid propellant systems. The first is called a "monopropellant," in which the liquid is a single compound, the molecule of which has the fuel and oxidizer components. For example, nitromethane is a monopropellant

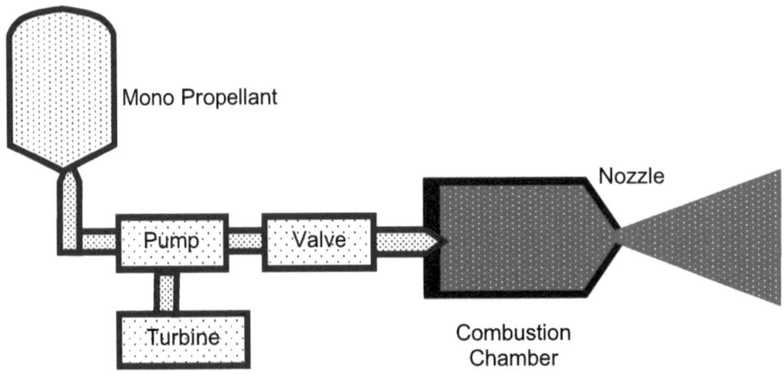

Figure 6.3
Schematic Representation of a Liquid Monopropellant Rocket Engine.

containing the fuel elements (carbon and hydrogen) and oxygen as the oxidizer. In any liquid rocket engines, the liquid propellant must be stored separately in a tank and needs to be pumped into the combustion chamber for operation. Figure 6.3 gives a schematic representation of a liquid monopropellant engine.

The second type of liquid rocket engine is based on a bipropellant system in which the oxidizer (in liquid form) and fuel (in liquid form) are separately stored in tanks. The oxidizer and fuel are pumped as per the required ratio into the rocket chamber for operation (Figure 6.4). This system obviously has more moving parts because of two separate flow systems; therefore, it has its own problems. Some of the well-known examples of bipropellant systems are as follows:

Oxidizer: Red fuming nitric acid (RFNA), hydrogen peroxide, and liquid oxygen
Fuel: Aromatic amines.

When we compare a solid propellant engine with a liquid propellant one, each has its advantages and disadvantages. For example, the design of a solid propellant grain is simpler and it does not have any additional moving parts (e.g., turbine/valve, etc.). However, once the solid propellant is ignited, it is difficult to stop or control the combustion whereas the flow of liquid oxidizer/fuel can be controlled.

In space programs and in advanced long-range ballistic missiles, the liquid propellant system is used either alone or along with a solid propellant system in different stages depending on the mission requirements.

The third type of rocket engine is called the "hybrid type" because it combines a solid (fuel/oxidizer) and a liquid (oxidizer/fuel). It is schematically shown in Figure 6.5. The liquid part (oxidizer; e.g., RFNA) is pumped into the rocket chamber containing the solid fuel (e.g., a polyurethane polymer). The hybrid propellant system has its own advantages and disadvantages of solid and liquid propellant systems.

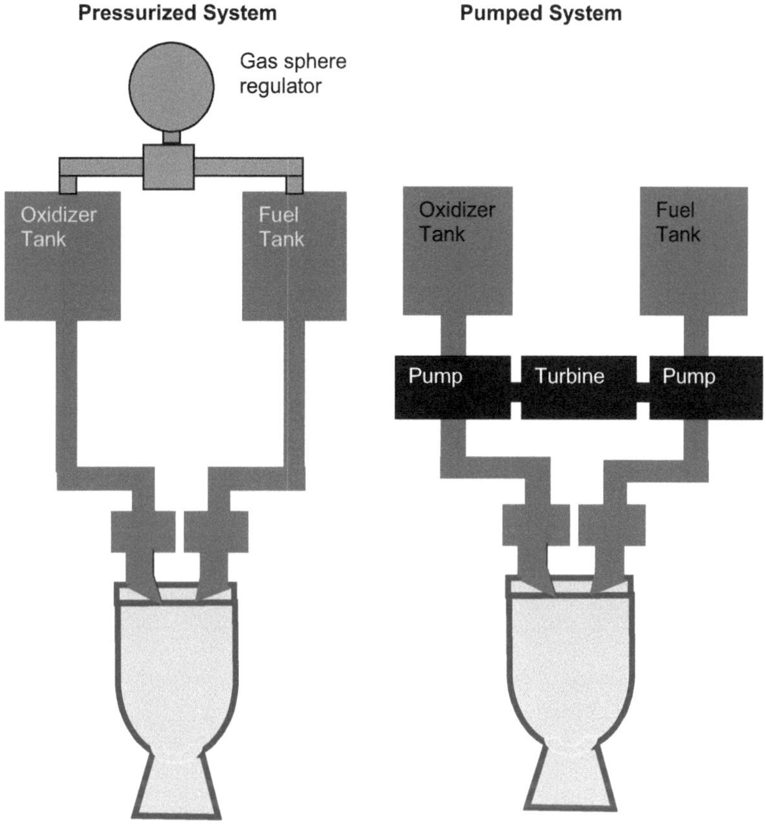

Figure 6.4
Schematic Representation of a Bipropellant Rocket Engine.

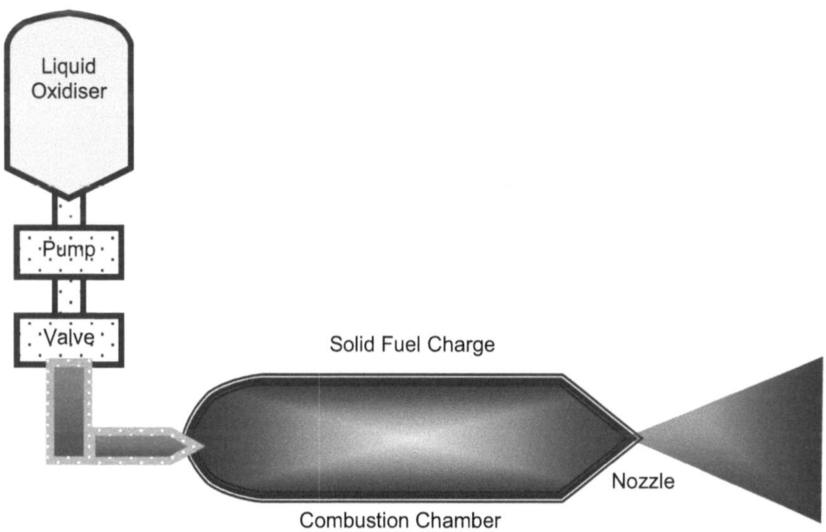

Figure 6.5
Schematic Representation of a Hybrid Liquid–Solid Rocket Engine.

6.3 Specific Impulse

Rocket designers have always been striving to achieve one goal—namely, to design a rocket

1. That can carry heavier payloads,
2. That can have longer ranges, and
3. In which the propellant consumption is minimal (analogous to fuel efficiency in automobiles).

Factors 1 and 2 demand that the total impulse developed by the rocket is quite high. Total impulse (I) is defined as $I = F \times t$ where F is the thrust developed by the rocket acting for a duration of time (t). In others words, factors 1 and 2 are directly proportional to $F \times t$. For complying with factor 3, the weight of the propellant consumed during the rocket flight (w) should be as little as possible. The term that considers these factors together to express the overall efficiency of a rocket propulsion system is called the "specific impulse" denoted by I_{sp}.

I_{sp} is accordingly expressed as

$$I_{sp} = \frac{F \times t}{w} \tag{6.4a}$$

This expression can also be written as

$$I_{sp} = \frac{\int F dt}{w} \tag{6.4b}$$

(or)

$$I_{sp} = \frac{F}{\dot{w}} \tag{6.4c}$$

where, $\dot{w}$ is the rate of consumption of propellant, being equal to $\frac{dw}{dt}$.

6.3.1 The Unit of I_{sp}

From Eqn (6.4a), we can see that F (i.e., thrust) and w (i.e., weight) have the same units—kilogram·meters per second squared (kg m s^{-2}; using SI units). Because they cancel out, only t remains. Therefore, specific impulse has the unit of seconds. For example, we can say that a given a propellant has an I_{sp} of 240 s.

6.3.2 I_{sp} and Exhaust Velocity of Gases

Let us consider a rocket cruising at a uniform velocity, and let the rocket function under the optimum nozzle expansion condition so that the second term in Eqn (6.3) is reduced to zero. If the exhaust gas velocity of the gases is v and the rate of loss of weight of the propellant (due to propellant burning) is $\dot{w}$, then the thrust (F) of the rocket, according to Newton's Second Law of Motion, is equal to the rate of change of momentum, which can be expressed as

$$F = \frac{d}{dt}(mv) = m\frac{dv}{dt} + v\frac{dm}{dt} = m\dot{v} + v\dot{m}$$

Because the rocket is moving with uniform velocity (i.e., $\dot{v} = 0$), in this case

$$F = \dot{m}v = \frac{\dot{w}v}{g} \tag{6.5}$$

(Because $m = w/g$).

Substituting this in Eqn (6.4c),

$$I_{sp} = \frac{\dot{w}v}{g} \times \frac{1}{\dot{w}} = \frac{v}{g}$$

$$I_{sp} = \frac{v}{g} \tag{6.6}$$

Therefore, I_{sp} is directly proportional to the exhaust velocity of gases (v).

Therefore, it is obvious that a propulsion scientist always endeavors to design his rocket—the hardware and the propellant—to achieve the highest possible value for the exhaust velocity (v).

Worked Example 6.1

A rocket develops a thrust of 10 tons by consuming 200 kg of propellant in 5 s. Calculate the specific impulse of the propellant used.

$$\text{The rate of propellant consumption} = \frac{200\text{ kg}}{5\text{ s}}$$

$$\dot{w} = 40\text{ kg s}^{-1}$$

$$I_{sp} = \frac{F}{\dot{w}} = \frac{10{,}000\text{ kg}}{40\text{ kg s}^{-1}} = 250\text{ s}$$

Why are we so specific about specific impulse?

It can be shown that the range of a rocket depends on the achievable *terminal velocity* (velocity of the rocket when the last gram of the propellant gets burnt), which again heavily depends on the I_{sp} of the propellant. I_{sp} plays a very vital role in the success of a mission. Every second gained in I_{sp} means very large gain in the range of a rocket. For instance, in the context of intercontinental ballistic missiles, an increase in the I_{sp} values by 1% and 5% increase their range by 7% and 45%, respectively. When a rocket is launched, its terminal velocity is severely limited because of two other forces: gravity and aerodynamic drag.

6.4 Thermochemistry of Rocket Propulsion

In the parlance of thermodynamics, a rocket can be called a "heat engine." The heat source is the high-temperature gaseous products obtained by the burning of the propellant. It uses part of that heat for the self-propulsive (or "useful") work, with the rest being wasted as heat loss by hot exhaust gases and by conduction of heat through walls of the rocket chamber. Therefore, rocket propulsion is the case of conversion of (a part of) the thermochemical energy of the propellant into the kinetic energy of the exhaust gases, a fact that is ultimately responsible for the rocket propulsion.

Let us designate that the initial heat content, pressure, volume, and temperature of the evolved gases during propellant deflagration be H_1, P_1, V_1, and T_1, respectively (Figure 6.6). The respective values for the exhaust gases can be assumed as H_2, P_2, V_2, and T_2. The change in heat content, $H_1 - H_2$, has been used to accelerate the exhaust gases to velocity v (i.e., assuming 100% conversion of thermal energy into kinetic energy of the exhaust gases). It can be written as

$H_1 - H_2 = \frac{1}{2}\frac{mv^2}{J}$ (i.e., kinetic energy of the gases)

(J = Joules constant = 4.18 J cal^{-1})

$$H_1 - H_2 = \frac{1}{2}\frac{w}{g}\frac{v^2}{J}$$

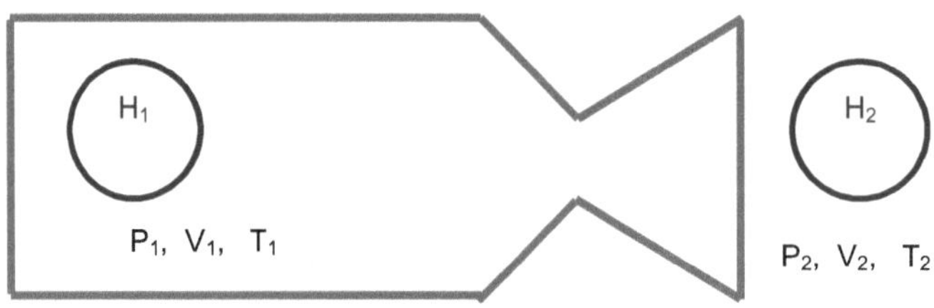

Figure 6.6
Change of Enthalpy and Other Parameters in Rocket Propulsion.

$$v = \sqrt{\frac{2gJ(H_1 - H_2)}{w}} \tag{6.7}$$

Assuming that the entire process is completely adiabatic (i.e., no heat is allowed to enter or leave the rocket motor system), it can be shown that

$$I_{sp} = \sqrt{\frac{2RT}{\overline{M}g}\left(\frac{\gamma}{\gamma - 1}\right)\left[1 - \left(\frac{P_e}{P_c}\right)^{\frac{\gamma-1}{\gamma}}\right]} \tag{6.8}$$

where $\overline{M}$ = the average molecular weight of the exhaust gases, T = the flame temperature of the propellant, P_e = the pressure of the exhaust gases, P_c = the chamber pressure, γ = the ratio of specific heats of the gases (average value), and R = is the universal gas constant.

For a given set of values for P_e and P_c, assuming that the value of γ has only a limited influence, it is seen that the I_{sp} of a rocket propellant mainly depends on the adiabatic flame temperature of the propellant and the average molecular weight of the exhaust gas products. The higher values of T and lower values of $\overline{M}$ favor higher I_{sp} values. Equation (6.8) can be written in a simplified manner as

$$I_{sp}\ \alpha\ \sqrt{\frac{RT}{\overline{M}}} \tag{6.9}$$

As mentioned earlier, the average molecular weight of exhaust gases and the flame temperature (isochoric in the case of gun propellants and isobaric for rocket propellants) of propellants greatly influence their performance. Although in the case of gun propellants their performance parameter, the force constant, is directly proportional to nRT_v, the performance parameter of a rocket propellant, I_{sp}, varies in direct proportion to the square root of nRT_p. (Note: $n = 1/\overline{M}$.) The theoretical I_{sp} calculated for a given propellant (as calculated in the above worked example) does not exactly agree with the I_{sp} measured when the a rocket is fired. It is because, in theory, we assume the performance of the rocket under ideal conditions, which deviate from the actual conditions of performance, is as follows:

1. The high-pressure gases inside of the rocket motor do not completely obey ideal gas laws.
2. The rocket motor does not ensure 100% thermal insulation, and the perfect adiabaticity of performance cannot be assumed.
3. The composition/homogeneity of the gases is not uniform through the entire length of flow.
4. The chemical equilibrium gets continuously shifted throughout the flow.
5. There are losses due to multidimensional flow. (In an ideal rocket, the flow is in one dimension only; i.e., along the x-axis.)

6. There is heat loss due to friction and other dissipative factors.

Because of these deviations, the *delivered* I_{sp} of a rocket is always less than the theoretical I_{sp}.

6.5 Some Vital Parameters in the Internal Ballistics of Rockets

It is proposed to highlight the importance of some parameters that greatly influence the rocket performance.

6.5.1 Linear Burning Rate

The linear burning rate (LBR; r) of a solid propellant grain decides the value of the mass burning rate, which is sometimes referred to as the "mass flow rate" ($\dot{m}$). You may recall we have related these two parameters as

$$\dot{m} = rA\rho,$$

where A and ρ refer to the surface area exposed for burning and the density of the propellant, respectively.

The parameters r and $\dot{m}$ greatly affect the performance time of the rocket, the pressure build-up pattern in the rocket chamber, etc. In the case of a solid rocket propellant grain, the value of r depends on the factors presented in the following subsections.

6.5.1.1 Chamber Pressure

In the case of DBRPs (based on NC and NG), r and the pressure chamber (P_c) are related as

$$r = bP_c^n \tag{6.10}$$

This equation is known as Vielle's law, in which n is the pressure exponent and b is the burning rate coefficient. It takes the logarithmic form as (see Figure 6.7)

$$\log r = \log b + n \log P_c \tag{6.11}$$

At a given temperature, a plot of log r against log P_c yields a straight line, the slope of which yields the value of n. This is an empirical law, generally obeyed by double-base propellants mostly in the range of conventional operating pressures (from 20 to 100 kg cm^{-2}). Composite propellants (i.e., propellants containing a polymeric fuel cum binder containing uniform dispersion of an oxidizer such as AP and metallic fuel such as aluminum powder) do not obey this law. Although a perfect model for composite propellant burning has not been developed so far, Summerfield's model

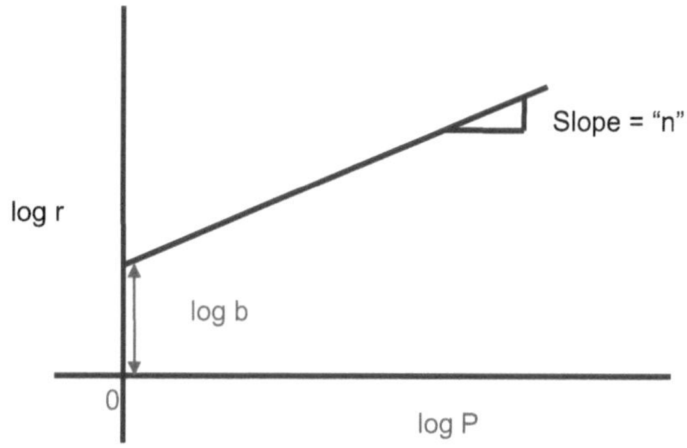

Figure 6.7
Plot of log r against log P for a Typical Double-Base Rocket Propellant.

has resulted in the following equation for composite propellants, which works reasonably well:

$$\frac{P}{r} = a + bP^{2/3} \tag{6.12}$$

with a and b being constants.

6.5.1.2 Temperature

The value of r increases with temperature. If r_1 and r_2 are the LBR values of a propellant at T_1 (in K) and T_2 (in K), respectively ($T_2 > T_1$), then the temperature sensitivity of burning rate at constant pressure, denoted as $(\pi_r)_P$, is given as

$$(\pi_r)_P = \frac{(\log r_2 - \log r_1)}{(T_2 - T_1)} \times 100$$

A rocket propellant designer always strives to keep the values of n and $(\pi_r)_P$ as low as possible. The higher these values, the greater are the chances for a catastrophic pressure build-up in a rocket motor.

6.5.1.3 Propellant Formulation

We have seen in the earlier chapters that if we formulate a propellant composition with high calorimetric value, then it results in higher flame temperature. It is natural to expect that in such cases, the heat transfer from the flame zone to the propellant surface will be faster, thereby increasing the r value of the propellant. In the case of composite propellants, apart from the aspects of the energetics of ingredients, the average particle size of the ingredients (oxidizer and metallic fuel) greatly affects the value of r for the same composition. The finer the particles, the greater is the value of r and vice versa.

The addition of burn rate catalysts also increases the value of r. For example, addition of salts/oxides of transition metals, such as Fe_2O_3 or $CuO \cdot Cr_2O_3$, as fine powder enhances the value of r. It is believed that electrons in the half-filled d orbitals of these transition metal atoms accelerate the decomposition of the AP (oxidizer) used in the composite propellants.

6.5.1.4 Erosive Burning

When high-velocity gases from the propellant erode the propellant surface, it results in faster heat transfer between the gas phase and solid phase, thereby increasing the value of r.

6.5.2 Characteristic Velocity

Referring back to the schematic representation of a rocket in Figure 6.1, let us ask: "What are the roles of propellants and the rocket nozzle in the rocket performance?" The first compartment (i.e., the chamber) ensures that the propellant burns as per the designed pressure-time profile and the high-pressure, high-temperature gases are ready to get into the nozzle to manifest their power. The total thermochemical energy of the propellant is being transformed into a high potential system ready for expansion through the nozzle. This thermochemical output is represented by the term "characteristic velocity" ($C*$; pronounced as "see star"), which represents the thermochemical potential of the propellant.

The nozzle then takes over. The high-pressure gases are initially compressed through the convergent portion of the nozzle and then expanded with enormous power through its divergent portion. The efficiency of nozzle expansion, which determines the value of the exhaust velocity of gases, is called the "thrust coefficient" (C_F), which is basically a thrust amplification factor. $C*$ is characteristic of a propellant in combination with the chamber and is independent of nozzle design. On the other hand, C_F is a thrust amplification factor and depends on the nozzle design. C_F is given by the equation

$$C_F = \frac{F}{P_C A_t} \tag{6.13}$$

The numerator in the right-hand term of the equation refers to the realized thrust whereas the denominator refers to the thrust experienced *at the throat* before it is amplified by the divergent section of the nozzle. Because the exhaust velocity (v) is determined by $C*$ and C_F, it can be written

$$v = C^* C_F \tag{6.14}$$

Therefore, $C*$ can be defined as the exhaust velocity of the gases when their pressure does not undergo any amplification by the nozzle ($C_F = 1$). However,

$$I_{sp} = \frac{v}{g} \quad \text{from Eqn (6.6)}$$

$$\text{Therefore,} \quad I_{sp} = \frac{C^* C_F}{g} \tag{6.15}$$

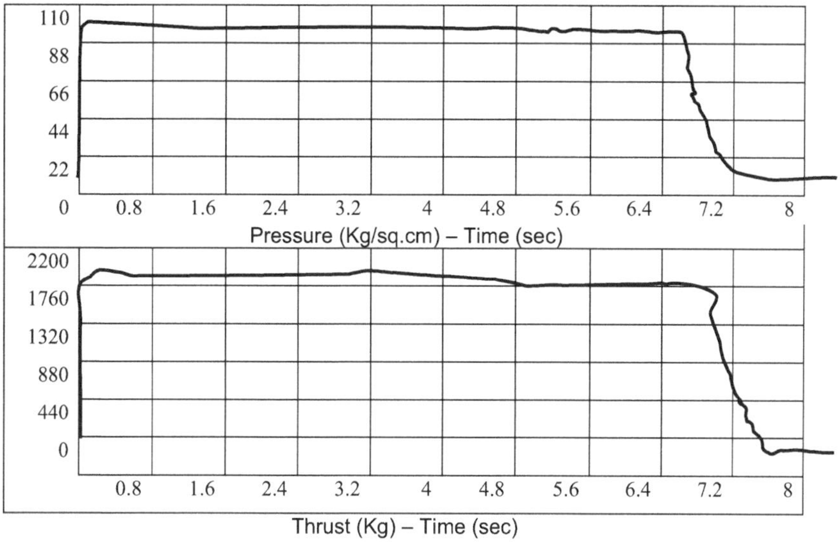

Figure 6.8
A Typical P–t and F–t Profile Obtained in a Static Rocket Firing.

$$I_{sp} = \frac{F}{\dot{w}} \quad \text{from Eqn (6.4c)}$$

$$\text{From Eqn (6.13),} \quad I_{sp} = \frac{C_F P_c A_t}{\dot{w}} \tag{6.16}$$

$$\text{From Eqns (6.15) and (6.16),} \quad C^* = \frac{g P_c A_t}{\dot{w}} \tag{6.17}$$

or

$$C^* = \frac{g A_t \int P dt}{W} \tag{6.18}$$

Using Eqn (6.18), C^* can be experimentally determined by finding out the area in a pressure–time curve (i.e., $\int Pdt$) by statically firing a rocket, knowing the values of W (weight of the propellant) and A_t (the cross-sectional area of the throat). A typical P–t and F–t profile curve obtained in a static rocket firing is shown in Figure 6.8. The areas under the P–t and F–t curves are obtained with great accuracy to calculate the parameters such as I_{sp} and C^*.

6.6. Design of a Rocket Propellant Grain

Depending on the requirements of a mission, such as payload, range, time of travel, etc., the ballistician finalizes certain basic parameters of propulsion such as (1) the total thrust needed, (2) the weight of the propellant (1 and 2 decide the I_{sp} of the propellant), (3) the action time (i.e., propellant burning time), (4) the density of the propellant, (5) the pressure index of the propellant, (6) the pressure at which the propellant should burn, (7) the A_e/A_t ratio, etc. Considering the mission requirements and the interdependence of many ballistic parameters, the ballistician has only a narrow choice. Once they finalize these, they turn to the propellant chemist and demand that the propellant must have characteristics such as

- An I_{sp} of x seconds,
- A C^* of y ms^{-1},
- An LBR of z mm s^{-1} at P kg cm^{-2},
- A pressure index in the range of n_1 and n_2,
- A density of ρ g cm^{-3}, etc.

It is now for the propellant chemist to use their expertise to formulate a composition that meets the demands of the ballistician. This is easier said than done. When they achieve one parameter (e.g., I_{sp}), some other parameters may start slipping away. For example, some of the higher energy (I_{sp}) versions of the solid rocket propellants may meet the demands of I_{sp} and C^* but may miserably fail in the requirement of pressure index. Or, they theoretically may feel confident about a particular formulation, but when they go to the process plant, they discover to their dismay that the composition is just not processable because the polymeric binder is too viscous to take the required solid loading of oxidizer and metallic fuel. It is like walking over a tight rope before the propellant chemist zeroes onto a certain formulation that satisfies the demands of the ballistician.

It is very often possible that the propellant chemist may not meet the demands of the ballistician exactly and there are slight variations. The ballistician then fine-tunes their design. Let us consider an example of a sustainer-type rocket propellant that burns from one end similar to a cigarette. The propellant chemist has finalized what they can offer and the ballistician works out the following parameters in sequence: (1) mass of the propellant, (2) the average burning surface, (3) the diameter of the grain, (4) the length of the grain, (5) the ratio of the burning area of the propellant at a given time to the area of the throat of the nozzle (called the K_N ratio), (6) the throat area (A_t), (7) the throat diameter, and (8) the area of the exit portion of the nozzle (A_e) on the basis of the A_e/A_t requirement.

Worked Example 6.2

How will you design a grain (cigarette-burning mode) of a rocket propellant considering the following requirements? (1) $I_{sp} = 200$ s, (2) $r = 0.5$ in s^{-1} at 1500 psi (which is the operating pressure), (3) $K_N = 400$ at 1500 psi, (4) density of the propellant $= 0.05$ lb in^{-3}, (5) $A_e/A_t = 10$, (6) thrust required $= 1000$ lb, and (7) burning time required $= 20$ s.

The sequence of calculation proceeds as follows:

1. Propellant weight : $\frac{F \times t_b}{I_{sp}} = \frac{1000 \times 20}{200} = 100$ lb
2. Grain length : $l = (r \times t_b) = 0.5 \times 20 = 10$ in
3. Grain volume : $\frac{\text{Weight}}{\text{Density}} = \frac{100}{0.05} = 2000$ in^3
4. Propellant diameter (D) : Volume $= \left[\frac{\pi D^2 l}{4}\right]$, Substituting the value for volume and length, $D = 16$ in
5. Propellant burning area : $A_b = \frac{\pi D^2}{4} = 200$ in^2

   (Cigarette-burning mode)

6. Area of the throat (A_t)

   Since $K_N = \frac{A_b}{A_t} = 400, \quad A_t = \frac{A_b}{400} = \frac{200}{400} = 0.5$ in^2

7. Area of the exit (A_e)

$$\text{Since } \frac{A_e}{A_t} \text{ should be 10,} \quad \frac{A_e}{0.5} = 10, \quad A_e = 5 \text{ in}^2$$

The above seven parameters are calculated by the ballistician so that they can accordingly design and fabricate the rocket motor to realize their requirements using the above propellant. This is probably the simplest example that can be given to make beginners understand the methodology of grain design. In actual practice, it is far more complicated, particularly while dealing with large grains with complex internal configurations. The propellant design might need modification by considering factors such as the extent of erosive burning, combustion instability, compatibility with the ignition system, etc.

6.7. Chemistry of Solid Rocket Propellants

6.7.1. Choices and Limitations

As already mentioned, the job of a propellant chemist is unenviable. They get the requirement from the ballistician, and then their rope walking commences. The chemist has to take into account several factors that the propellant should satisfy, and the major ones are presented in the following subsections.

6.7.1.1 The Energetics

It is precisely the specific impulse (I_{sp}), the index of energy of any rocket propellant, that needs major consideration. I_{sp} varies depending on the nature of propellant. We have seen (Eqn (6.8)) that for a given chamber and exit pressure, I_{sp} mainly depends on the flame temperature and the average molecular weight of the products. We have seen in detail in Chapter 2 that the flame temperature depends on the calorimetric value (heat of explosion)

Table 6.1: Rocket propellant formulations with calorimetric value and theoretical I_{sp} values (values of P_c and P_e taken as 70 and 1 kg cm^{-2}, respectively).

Composition	Calorimetric Value (cal g^{-1})	I_{sp} (Theoretical; s)
Cast double base	800–1000	200–220
Extruded double base	800–1050	200–220
Composite	1000–1200	Up to 245
CMDB	900–1300	Up to 260
Nitramine double base	1000–1200	Up to 235

CMDB, composite modified double-base.

of the propellant whereas the average molecular weight ($\overline{M}$) of the product gases depends on the relative amounts of C, H, N, O, and other elements. If we compare DBRPs and composite rocket propellants (CRPs), it is seen that the I_{sp} of CRPs is more than that of DBRPs (see Table 6.1).

Although the average molecular weight of gaseous products is more in the case of CRPs because of the presence of chlorine, mainly as hydrogen chloride (molecular weight = 36.5 mole^{-1}; chlorine originating from the oxidizer, AP), the higher heat output due to the highly exothermic oxidation of aluminum (fuel) more than compensates for the molecular weight factor. On the other hand, composite modified double-base (CMDB) rocket propellants, which are an intelligent combination of CRP and DBRP (CMDB propellant uses an energetic polymer matrix based on a double base, i.e., NC and NG, in which AP and aluminum are incorporated), exhibit I_{sp} much higher than even those of CRPs. The major drawback of CMDB rocket propellants is their sensitivity to mechanical initiation due to the presence of NC and NG.

6.7.1.2. Burn Rate and Other Ballistic Parameters

A solid rocket propellant should burn at a specified LBR (r) at its operating pressure. The propellant chemist realizes that r depends on various factors, such as

1. The heat of explosion (to which it is directly proportional);
2. The presence/absence of a catalyst (e.g., Fe_2O_3 is used as a burn rate catalyst in some cases because it is believed to catalyze the decomposition of AP through an electron transfer mechanism setting out a host of free radical and ionic species that catalytically pyrolyze the polymer matrix);
3. The particle size and its distribution in the case of the oxidizer particles in CRPs and CMDB rocket propellants (generally, the lower the average particle size, the higher the specific surface area resulting in higher mass burn rate);
4. The presence of heat conducting substances (e.g., addition of carbon black); and
5. Erosive burning conditions.

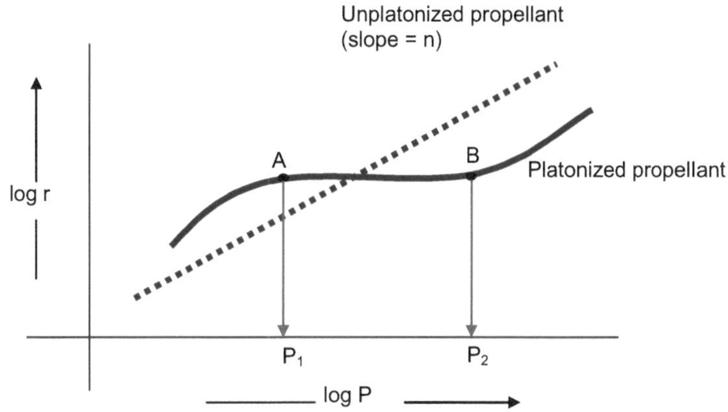

Figure 6.9
Plateau Effect.

Apart from burn rate catalysts, in some cases, certain substances need to be added in the propellant composition to ensure that the value of *r* does not change between certain pressure ranges. This is called a "plateau" condition, and the substances added for this purpose are called "platonizers" (see Figure 6.9).

Addition of platonizing agents, such as basic lead stearate in DBRPs, has been found to be effective in achieving a pressure independence of the burn rate between certain pressure ranges (P_1 to P_2). The value of the pressure exponent n is nearly zero in this region. Substances such as basic lead stearate, basic lead salicylate, etc., have been successfully used to achieve platonization in DBRPs.

6.7.1.3. Processability

While taking care of the energetics, the propellant formulator has to consider the processability of what he intends to formulate. He will have to carefully analyze the interdependence of various factors of processability and choose the most optimal formulation. Let us illustrate this with an example of a CRP formulation.

A CRP contains an oxidizer (mostly AP) and a metallic powder (e.g., aluminum powder) dispersed in a polymeric matrix (e.g., a polyurethane matrix that plays the dual role of binding AP and aluminum, thereby structural integrity to the propellant grain and a fuel). The formulation also contains smaller percentages of other ingredients such as a plasticizer, process aid, burn rate catalyst, etc. A typical composition is as follows:

1. AP = 68% (2:1 mixture of coarse AP (~250 μm) and fine AP (~10 μm)),
2. Aluminum = 17%,
3. Polymer = 15% (polyurethane, based on hydroxyl terminated polybutadiene (HTPB)), and
4. One part of Fe_2O_3 (burn rate catalyst).

The major steps involved are as follows:

1. Ingredient preparation
   a. Drying of AP and blending of the coarse and fine varieties (it is called "bimodal AP." The purpose of blending coarse and fine AP is to achieve maximum loading of AP in the thick viscous "prepolymer"-like HTPB before the prepolymer is cured. Remember the mason mixing fine cement with sand particles of various sizes while making concrete mixture? Such a multimodal mixing ensures that the interstices between bigger particles are filled by smaller particles ensuring maximum space utilization).
   b. Drying of aluminum powder.
   c. Drying of HTPB, plasticizers, etc.
2. Mixing

   All of the above ingredients are mixed in a mixer.

3. Addition of curative

   The addition of curative sets in the curing reaction. In our example, addition of toluene di-isocyanate (TDI) starts the curing reaction (the NCO groups of TDI react with the terminal OH groups of HTPB forming the urethane, NH.COOR linkage between HTPB prepolymer molecules), and the slurry mix becomes more viscous.

4. Casting

   The slurry is cast into an inner-lined rocket motor fitted with a mandrel. (Note: The slurry should be poured/cast into the motor before its viscosity increases beyond a certain level.)

5. Curing

   The rocket motor into which the propellant slurry has been cast is kept in an oven, the temperature of which may be approximately 70 °C for approximately 7 days.

6. Mandrel removal after removal of the rocket motor from oven and cooling. The propellant curing process is over and the mandrel is carefully removed. After necessary inspection (e.g., X-ray inspection to ensure the absence of defects such as cracks and voids in the solid grain), the motor is ready for firing after further assembly.

Let us say that the above formulation realizes the following performance parameters:

$I_{sp} \cong 245$ s, burn rate $= 10$ mm s^{-1} at 70 kg cm^{-2}

If the propellant chemist is asked to modify the composition so as to increase the I_{sp} to 250 s without affecting the burn rate, then what options are left to them and with what consequences?

Option 1

Because I_{sp} is directly proportional to flame temperature and heat of explosion, they can go in for higher loading of AP and aluminum so that the solid loading goes from 85% to 87%.

Consequence 1

This 2% increase in solid loading is too much for the HTPB prepolymer to take. It becomes difficult to mix. Even if they manage to mix, the viscosity increase of the mix after the addition of curative is too fast for smooth casting. The propellant made out of this mix is very likely to contain many voids that are unacceptable.

Consequence 2

Higher AP and aluminum means higher flame temperature, which will increase the burn rate beyond 10 mm s^{-1}, which also is unacceptable.

Consequence 3

A lower percentage of the polymeric matrix in the final propellant will adversely affect the mechanical property of the propellant, leading to a lower percentage of elongation and a lower glass transition temperature.

Option 2

Increase the solid loading by 2% but compensate it by decreasing the ratio of finer AP so that viscosity increase can be taken care of.

Consequence 1

When the percentage of fine AP decreases (or that of coarse AP increases), the burn rate of the final propellant will decrease, which is not acceptable.

Consequence 2

A large increase in coarse AP percentage will also affect the mechanical property of the final propellant by reducing its tensile strength.

Option 3

Increase the solid loading by 2% but use the prepolymer HTPB of lower viscosity to take care of the viscosity build-up.

Consequence 1

Lower viscosity means lesser chain length/molecular weight of HTPB, and this will result in poor mechanical property of the final grain.

Consequence 2

A higher burn rate.

The above is just one example of the complex interplay of various parameters of formulation viz-a-viz the processability. The propellant chemist needs to blend chemistry and experience to solve such problems for which there are no quick-fix solutions.

6.7.1.4. Mechanical Properties

A rocket propellant grain has to withstand various mechanical stresses right from the time it is made until it is used. At various stages such as transportation, storage, assembly, and actual flight it undergoes tensile and compressive loads, shocks, vibration, high G values, etc., and if the mechanical properties of the grain are too poor to withstand such stresses, then it will result in the formation of abnormalities such as cracks, which are highly disastrous. In general, a propellant grain, which is free-standing and gets loaded to a motor after inhibition, requires a high tensile strength. In the case of case-bonded propellant charges (i.e., in which the propellant is directly cast into lined motors in situ), the grain should be able to take high compressive load and should therefore have higher elongation.

6.7.1.5. Storage Stability/Life

The rocket propellants, particularly meant for military use, are stored in assembled motors for a long period under varying conditions of temperature and humidity. The propellant chemist has to carefully analyze the compatibility among various ingredients that are used in propellant processing. For instance, an incompatible ingredient may accelerate the breakdown of the polymer matrix in a propellant, resulting in the development of cracks. There are well-established methods, such as several surveillance tests and thermal analyses, which can help in assessing such incompatibilities.

6.7.1.6 Safety and Environment: Cause for Concern

Safety is the most important factor that should be foremost on the mind of any HEM chemist. They are handling materials that have all of the three types of risks: explosion, fire, and toxicity. It is a known fact that under extreme conditions, such as undue confinement, a propellant or many of its ingredients can violently detonate. Fire risk is always there with any type of propellant. Quite a few chemicals used in propellant processing are carcinogenic and mutagenic. When the propellant is finally processed, it must be reasonably insensitive to impact, friction, and static discharge. There is no relevance in formulating a high-energy propellant that is quite dangerous to handle.

All over the world, the propellant scientists and technologists are exploring the possibility of going in for ecofriendly or "green" propellants and propellant ingredients. For example, despite many of its attractive properties (e.g., low cost, better energetics, and stability), AP is found to be ecologically detrimental when used in several tons. Large quantities of chlorine-related products emitted in the upper atmosphere when AP-based propellants burn cause environmental problems such as acid rain and ozone depletion. Efforts are on to replace AP with new, ecofriendly (chlorine-free) oxidizers such as ammonium dinitramide (ADN) and hydrazinium nitroformate (HNF). The above are the major six factors that the propellant chemist must keep in mind while formulating a propellant for a given mission, apart from considerations such as cost and availability of raw materials.

6.8 Future of Rocket Propellants

The progress in the field of rocket propellants has been painfully slow despite the enormous amount of research going on all over the world. This is primarily due to the conflicting conditions and requirements that confront propellant chemists, such as energetics, cost, safety, stability, and environmental friendliness. When a candidate propellant ingredient is synthesized, it is very exhaustively tested for all of these criteria before it can be introduced in a rocket propellant formulation. For example, it took several decades to replace the good old polyvinylchloride-based plastisol propellants with today's workhorse propellant that is based on HTPB. Many later versions of binders such as glycidyl azide polymer and oxetane-based polymers and copolymers containing energetic functional groups such as nitro, nitrato, and azido groups have their own disadvantages and still HTPB is reigning supreme. Despite the loud cries against the ecological impact of AP, it is still the most used oxidizer because of its many attractive properties. The alterative candidates have certain serious disadvantages. For example, HNF is still not safe enough for large-scale processing because of its high sensitivity to friction. ADN is not attractively energetic, and its high hygroscopicity poses problems for processing. The same argument applies when we search for better metallic fuels to replace aluminum. Beryllium gives more energy on oxidation, but the products are unacceptably toxic. Lithium is less energetic. On combustion, boron gives problematic products. Much research is going on all over the world in this direction, and we hope that we discover better oxidizers, fuels, plasticizers, burn rate catalysts, etc., in the foreseeable future so that we can aim for longer ranges and higher payloads in tomorrow's rockets.

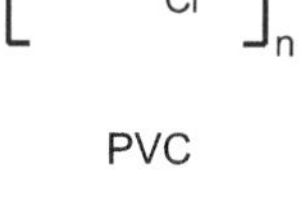

PVC

HTPB

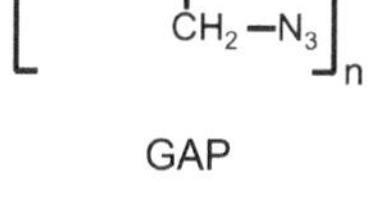

GAP

NH_4ClO_4

AP

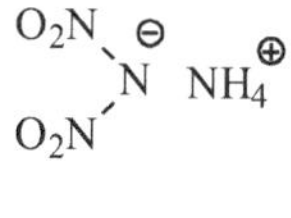

ADN

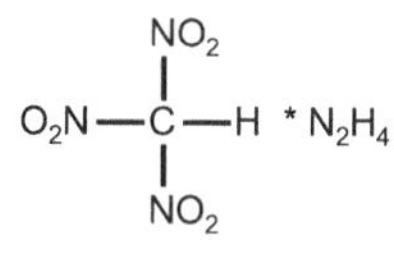

HNF

PVC: Polyvinylchloride

HTPB: Hydroxyl terminated polybutadiene

GAP: Glycidyl azide polymer

AP: Ammonium perchlorate

ADN: Ammonium dinitramide

HNF: Hydrazinium nitroformate.

Suggested Reading

[1] R. Meyer, J. Kohler, Explosives, VCH Publishers, Germany, 1993 (Encyclopaedia – handy for referencing).
[2] T. Urbanski, Chemistry and Technology of Explosives, vol. 1–4, Pergamon Press, Oxford, New York, 1983.
[3] A. Bailey, S.G. Murray, Explosives, Propellants and Pyrotechnics, Pergamon Press, Oxford, New York, 1988.
[4] B. Siegel, L. Schieler, Energetics of Propellant Chemistry, John Wiley & Sons. Inc., New York, 1964.
[5] S.F. Sarner, Propellant Chemistry, Reinhold Publishing Corporation, New York, 1966.
[6] S. Fordham, High Explosives and Propellants, Pergamon Press, Oxford, New York, 1980.
[7] J.P. Agarwal, High Energy Materials, Propellants, Explosives and Pyrotechnics, Wiley, 2010.
[8] N. Kubota, Propellants and Explosives Thermochemical Aspects of Combustion, 2007.

Questions

1. What are the two major parts of a solid rocket motor?
2. What is the role of a CD nozzle in a rocket motor?
3. Can you explain, using the thrust equation, why the thrust experienced by a rocket is maximal while it traverses through vacuum?
4. What are the relative merits and demerits of solid and liquid rocket engines?
5. Explain why the unit of specific impulse is expressed in seconds and how it is related to the exhaust velocity of gases.
6. Calculate the weight of a solid rocket propellant ($I_{sp} = 210$ s) that should be loaded in a rocket motor to produce a thrust of 6 tons. The propellant burns for 4 s (Answer: 114.3 kg).
7. What are the two major characteristics that decide the value of its specific impulse?
8. What are the factors that reduce the actual (realized) I_{sp} of the propellant compared with the theoretical I_{sp} calculated?
9. What is Vielle's law and why must a propellant chemist be worried about the value of n, the pressure index?
10. What is the significance of C^*?
11. Why is the job of a rocket propellant chemist similar to walking on a rope?

12. Why are CMDB propellants more energetic than the composite and double-base propellants?
13. What are the major steps involved in processing composite propellants?
14. Case-bonded rocket propellants should have high compressive strength and elongation. Why?
15. Name some of the potential candidates for polymeric binders, fuels, and oxidizers for use in solid rocket propellants.

CHAPTER 7

High Energy Materials in Pyrotechnics

7.1 Introduction

Common man understands rather simplistically that "pyrotechnics" means fireworks. It is generally known that the first people to develop fireworks were the Chinese more than 1000 years ago. The Chinese were experts in the field of pyrotechnics, and as early as the tenth century they had developed rockets. As mentioned in Chapter 1, the English scientist Roger Bacon made a quantitative study of gunpowder during the thirteenth century, and the use of gunpowder as a propellant for cannons was prevalent in Europe in the fourteenth century. When the application of pyrotechnics (meaning the art of making and using fireworks) underwent a transition from civilian use to military use, enormous efforts to search for suitable chemicals and innovation in formulation and processing led to significant progress in the field of pyrotechnics.

7.2 Applications

Let us consider the following scenario. A multistage rocket takes off either for a military or for a space mission. First-stage propulsion is only possible if the propellant is suitably ignited by an igniter, which is basically a pyrotechnic composition. In the case of one-stage, small-size rockets, it may be a cartridge containing gunpowder of certain charge weight and granular size. The higher version of the propulsion may engage an igniter containing a pyrotechnic mixture of magnesium, potassium nitrate (KNO_3), and a binder.

The very success of the mission depends on the correct formulation, charge weight, and granular size of the igniter. Examples include the following:

1. An igniter composition for a double-base rocket propellant is mainly based on gunpowder. The design of the igniter (quantity, particle size, shape of the container containing the igniter) depends on the propellant characteristics, and many trials need to be conducted to ensure the propellant-igniter matching. The gunpowder is housed in a cambric cloth bag and placed in the port (annular) area of the propellant grain.
2. A common igniter composition for a composite rocket propellant consists of an oxidizer, a metallic fuel, and a binder (e.g., boron/KNO_3/binder). The igniter composition is housed in a metallic tube, which easily ruptures on initiation of the igniter, spreading the flame throughout the port area of the propellant. There often

Demystifying Explosives: Concepts in High Energy Materials. http://dx.doi.org/10.1016/B978-0-12-801576-6.00007-0

arises a need to introduce a delay of a certain fixed period (varying from milliseconds to seconds) to actuate a device, which may be a detonator or a propulsive system. A delay cartridge containing certain pyrotechnic composition helps to achieve this. The formulation of the composition must be extremely precise and calls for thorough knowledge and experimental trials involving various pyrotechnic ingredients. A typical delay composition follows is barium chromate ($BaCrO_4$)/antimony trisulfide/potassium perchlorate.

In certain war scenarios, it becomes necessary to destroy the enemy targets by sheer heat rather than explosion. Incendiary ammunitions used for this purpose are basically pyrotechnic compositions and use pyrophoric (ignites when comes into contact with air) ingredients such as zirconium. A typical incendiary composition based on zirconium is zirconium/crepe rubber.

During night warfare, it often becomes necessary to illuminate the enemy territory using illuminating pyrotechnic compositions with a specified illuminating time and intensity to the tune of several thousands or millions of candelas. A typical illuminating composition is magnesium/sodium nitrate ($NaNO_3$)/resin (binder).

Signaling plays a crucial role in any warfare and during emergencies in peace time. (Signaling pyro compositions were launched in large numbers when the *Titanic* was sinking during night more than a century ago.) Pyrotechnic compositions with varying signaling implications were developed a long time ago and are still in use. Typical signaling compositions include

- Magnesium/strontium nitrate (Sr(NO3)2)/resin
- Magnesium/$NaNO_3$/resin

Pyrotechnic compositions are also used to track a target in air. They are also known as "tracer compositions." A typical tracer composition is magnesium/$Sr(NO_3)_2$/$NaNO_3$/resin.

In tactical warfare, decoy flares are still being used to decoy the heat-seeking enemy missiles and protect the aircraft from which flares are launched to divert the missiles. The pyrotechnic composition of the flares fakes the signals (mainly infrared (IR)-based signals) of the aircraft. A typical decoy composition is magnesium/Teflon/Viton.

Several pyrotechnic compositions produce smoke for visual obscuration (some special compositions also produce smoke that is impervious to IR radiation) or signaling (using smokes of specific colors). An example is red phosphorous/KNO_3/resin.

It is interesting to note that certain pyrotechnic compositions have been developed either for signaling or distraction purposes. For example, a certain composition may create the sound of an aircraft to confuse the enemy.

7.3 Basic Principles of Pyrotechnics

7.3.1 The Chemical Components of Pyrotechnics

The basic chemical ingredients of pyrotechnics are an oxidizer, a fuel, a binder (in most of the cases), and often a chemical or mixture of chemicals added to give various effects as seen in Section 7.2. At times, the term "pyrotechnics" is loosely used even in the case in which there is no burning involved. For instance, a smoke composition to produce smoke may involve a compound such as titanium tetrachloride, which on hydrolysis gives intense smoke, and such a composition is also categorized under pyrotechnics.

7.3.1.1 Oxidizers

Pyrotechnic reactions are mostly solid–solid reactions. All oxidizers used are solid ones in the form of fine powder, and the particle size of the oxidizers to be used should strictly fall within a specified range. Most of the oxidizers are salts of metals such as chlorates (e.g., potassium chlorate), chromates (e.g., $BaCrO_4$), dichromates (e.g., potassium dichromate), nitrates (e.g., KNO_3), and oxides (e.g., barium peroxide). All of these salts evolve oxygen during decomposition, which is used to oxidize the fuels. Halogens are known to be good oxidizing agents; therefore, compounds such as Teflon (C_2F_4 polymer) are used effectively in certain pyrotechnic compositions as oxidizers. While choosing an oxidizer for certain pyrotechnic compositions, the following factors should be carefully considered:

1. Energetically, the oxidizer must have an acceptable heat of decomposition. If the value is too high, then the high exothermicity may result in the explosion of the pyrotechnic composition. If it is too low, then the low heat output may not even ignite the pyrotechnic composition or the rate of burning may be quite low.
2. Most of the oxidizer salts used contain alkali metals (e.g., KNO_3) or alkaline earth metals (e.g., $Sr(NO_3)_2$) as cations because these metals are poor electron acceptors (rather excellent electron donors); hence, they will not react with metallic fuels such as magnesium or aluminum. For example, we can never expect a reaction such as

$$2Na^+ + Mg \rightarrow 2Na + Mg^{2+}.$$

3. Because ingress of even a very small amount of moisture content plays havoc with the performance of pyrotechnic compositions (leading in extreme cases to fire or explosion), the oxidizer must have very low hygroscopicity. The strict adherence to humidity control during the processing of pyrotechnic compositions is due to the same reason.
4. The chosen oxidizers should be low in toxicity and should not be too sensitive to friction and impact to ensure safety of personnel during processing, transport, and storage.

7.3.1.2 Fuels

The fuels used in pyrotechnics are powdered elements (either metals or nonmetals) that provide sufficient energy on oxidation. While choosing a fuel–oxidizer combination, one should carefully assess the quantum of heat output (that determines the flame temperature) and the nature of the products. Metallic fuels are used where there is a need for high heat output and hence high flame temperature. For example, in illuminating compositions, a high flame temperature is a must to ensure intense light emission. Magnesium is one of the favorite candidates in many illuminating compositions because the heat of oxidation of magnesium is very high, resulting in the formation of incandescent magnesium oxide (MgO) particles that help in the highly intense light output. Conversely, metals such as magnesium cannot be used in compositions in which heat output has to be low, as in colored smoke compositions using organic dyes. High heat output will decompose the dyes, defeating the very purpose of the colored smoke production. In such a composition, low-calorie fuels such as sugars can be used.

7.3.1.3 Binders

We have seen [refer 6.7.1.3] that binders play dual role in processing composite rocket propellants. They not only give structural integrity to the finished propellant but also act as a source of organic fuel during propellant burning. Binders used in pyrotechnic compositions (both natural binders such as shellac, beeswax, and artificial ones such as polyvinylchloride and epoxy resins) play the following roles:

1. consolidate the composition by increasing the cohesive forces between all of the particles.
2. Binders coat and protect reactive ingredients such as metal powders, which otherwise may easily be oxidized by atmospheric oxygen.
3. Binders reduce the sensitivity of the composition to impact and other sources of stimuli.
4. In some cases, binders modify the burning rate of the final composition.

The binder chosen must be neutral (neither acidic nor basic) and nonhygroscopic to prevent any problems during the production of the pyrotechnic composition or storage. For example, a water-based binder is bound to create problems where magnesium is used because the latter is very reactive with water. Also, the binder should result in the proper consolidation/structural integrity of the final product.

7.3.1.4 Other Ingredients

Retardants are chemicals that are added to certain pyrotechnic compositions to reduce the burning rate below a desired level. These retardants are basically chemicals that absorb heat (endothermic) for their decomposition, such as carbonates, bicarbonates, and oxalates of alkali and alkaline earth metals.

For instance, calcium oxalate (monohydrate) added to the composition endothermically decomposes as follows:

$$Ca(C_2O_4)\cdot H_2O \overset{\text{Heat}}{\rightarrow} CaO + CO + CO_2 + H_2O$$

Because the oxalate absorbs heat during this decomposition, it produces the cooling effect and thereby decreases the flame temperature and hence the burning rate of the pyrotechnic composition.

7.3.2 Factors Affecting the Performance of Pyrotechnics

Pyrotechnic reactions are basically solid–solid reactions, and the performance of a pyrotechnic composition largely depends on certain parameters concerning those solids (powders), whether they are oxidizers, fuels, inert fillers, etc. Some of these parameters are presented in the following subsections.

7.3.2.1 Stoichiometry

The reactants involved in a pyrotechnic reaction should be taken in the stoichiometric ratio to achieve a balanced reaction. This will ensure the maximum output of heat and the highest rate of burning. On the other hand, if excess of either fuel or oxidizer is taken, then the net heat output per gram of the composition will be lower than what is required.

7.3.2.2 Particle Size

The importance of the particle size of ingredients in determining the rate of burning of a high-energy material composition has been already dealt with in earlier chapters when we discussed linear and mass burning rates. In the case of pyrotechnic performance, which is a solid–solid reaction, this factor becomes extremely important. The average particle size of a compound (roughly assuming a spherical nature of each particle) determines the specific surface area (expressed as m^2/kg or cm^2/g). It is the specific surface area and the thoroughness of mixing the ingredients that will determine how "intimate" the contact between an oxidizer and a fuel (or any other ingredients) is in a pyrotechnic composition. Therefore, this calls for a serious quality-control check at the time of ingredient preparation with respect to the adherence to particle size limits as specified for a given pyrotechnic composition.

7.3.2.3 Avoiding Material Degradation during Storage

Almost all of the pyrotechnic compositions involve an intimate mixture of finely divided metals, fine powders of oxidizers, and other ingredients. Because of the high specific surface area involved, these compositions are highly vulnerable to degradation during storage. For example, finely divided magnesium powder is quite susceptible to oxidation by atmospheric oxygen, and the formation of any MgO coating will hamper the performance of the composition. To obviate such a problem, magnesium powder is coated with inert materials, such as lacquers and varnishes, before it is incorporated in the composition. Some of the oxidizers such as $NaNO_3$ are known for their hygroscopicity and, on storage, the ingress of moisture and the subsequent moistening or even the dissolution of the oxidizer component in the composition will severely hamper the satisfactory performance of the pyrotechnic composition. Therefore, it is imperative that the finished product should be hermetically sealed to prevent any ingress of moisture.

7.3.3 Safety Aspects Involving Pyrotechnics

Following strict safety precautions becomes mandatory at every stage when it concerns pyrotechnics, including at the design/formulation of composition, the preparation of ingredients, the processing of the final composition, packing, transport, and storage. The high level of hazard connected with pyrotechnic compositions is due to two factors: (1) the ingredients are very sensitive either individually (e.g., pyrophoric Zr) or in combination (e.g., thermite composition such as $Al + Fe_2O_3$) and (2) the exposed surface area of ingredients is very high because of the low particle size, at times going down even to the submicron level in certain compositions. In some cases, the hazard is enhanced because of the gritty or sharp-edged nature of some crystalline powders, in which case due care must be taken during processing.

Before embarking on any new composition, a thorough literature survey and analysis of the Material Safety Data Sheet should be performed to evaluate the hazards (fire, explosion, and toxicity hazards) of the proposed ingredients. Even more important is the careful study of the compatibility of the ingredients proposed to be incorporated in the composition. Many ingredients, although harmless individually, may result in disasters when mixed with others without taking adequate precautions. Some examples are as follows:

1. Chlorates are highly incompatible with sulfur and phosphorous (the slow formation of the acids of sulfur and phosphorus on storage in the presence of moisture and their subsequent reaction with chlorates result in highly unstable and explosive chloric acid) as well as with carbonaceous and ammonium compounds.
2. Very fine ammonium perchlorate or ammonium nitrate can be dangerously sensitive to impact in the presence of carbonaceous impurities.
3. Even traces of water can be very dangerous when it comes into contact with mixtures containing finely divided zirconium, titanium, magnesium, zinc, or aluminum.

By and large, most of the pyrotechnic compositions are sensitive to friction, impact, flame, and static discharge. When preparing large quantities, operations such as mixing are done under remote control. Although hand-mixing is done for smaller quantities, it is mandatory to use safety equipment/infrastructure such as conductive mats, conductive gloves, etc., that are all connected to a properly working static discharge system. This will ensure that no static charge is allowed to remain in the vicinity of the composition being mixed. We must remember that certain compositions can be ignited with a static discharge of a potential as low as a few millivolts. Because the development of static charge is closely related to the humidity level in the processing room (lower humidity favoring it), humidifiers should be in operation during processing to maintain the specified range of relative humidity.

Many accidents have been reported during the waste disposal of pyrotechnic stores. Proper standard operating procedures should be formulated and strictly followed for each type of pyrotechnic composition when it comes to its disposal.

7.4 Conclusion

Pyrotechnics have come a long way over centuries, from gunpowder to sophisticated pyrodevices used in various applications for defense as well as space missions. The very success of such missions heavily depends on the reliable and satisfactory performance of the pyrotechnic component in the explosive train involved. Although it may be commonly said that "pyrotechnics making is an art," the fact is that this field is a multidisciplinary one involving solid state chemistry and engineering. Despite their usefulness, it should be remembered that pyrotechnics are very sensitive to mechanical impacts, heat/fire, and static discharge and can result in disasters if the safety rules are not respected.

Suggested Reading

[1] J.A. Conkling, C. Mocella, Chemistry of Pyrotechnics: Basic Principles and Theory, second ed., 1947.
[2] Pyrotechnic chemistry, Journal of Pyrotechnics (2005). Pyrotechnic series.
[3] J. Akhavan, The Chemistry of Explosives, third ed., Royal Society of Chemistry, 2011.
[4] J.P. Agarwal, High Energy Materials, Propellants, Explosives and Pyrotechnics, Wiley, 2010.
[5] R. Meyer, J. Köhler, A. Homburg, Explosives, 2007.
[6] N. Kubota, Propellants and Explosives Thermochemical Aspects of Combustion, 2007.
[7] U. Teipel, Energetic Materials Particle Processing and Characterization, 2005.
[8] M. Hattwig, H. Steen, Handbook of Explosion Prevention and Protection, 2004.

Questions

1. Which is the oldest pyrotechnic composition known to man?
2. What factors of gunpowder are important when it is to be used as an igniter for a rocket propellant?

3. A typical igniter composition used in a composite rocket propellant is given as boron/KNO_3/plasticized ethyl cellulose. What is the role of each of these ingredients?
4. What is meant by the term "pyrophoric"? Give an example of a pyrophoric substance.
5. Teflon is a well-known polymer and does not contain oxygen in its molecule. How then is it used as an oxidizer?
6. Most of the oxidizer salts used in pyrotechnic compositions contain either alkali or alkaline earth metals. Why?
7. Why can we not use a high caloric value composition for producing color smokes?
8. What is specific surface area and what are its units? Why is this parameter very critical when formulating pyrotechnic compositions?
9. Why do we prefer to coat magnesium powder with lacquers or varnishes before we use it in pyrotechnic compositions?
10. Why are lower humidity levels dangerous when processing pyrotechnic compositions?

CHAPTER 8

HEMs: Concerns of Safety

8.1 Introduction

Do you know a strange fact? Although explosives are dangerous and feared substances, the explosives industry does not figure in the top ten among the most accident-prone industries or professions in the world (coal mining and steel industries are at the top of the list). This is obvious because those who deal with the explosives know that they deal with the explosives! A whole range of precautions are taken, Standard Operating Procedures (SOPs) are followed, and clearly written-down DO's and DON'Ts are observed at every stage of explosives processing all over the world. Nevertheless, accidents, some of them disastrous, still keep occurring sporadically, indicating that some lapses must have occurred either due to ignorance or negligence. Remember, in the field of HEMs, it is safety and safety alone that is the priority, and the rest of the objectives, like project success, cost, etc., come later. The intention of this chapter is to give the readers a gist of the vital and salient points concerning various aspects of HEM safety.

8.2 Nature of Hazards

In the earlier chapters we have seen that HEMs can result either in detonation (creating destructive shock waves) accompanied by blast or deflagration, depending on the circumstances that they are subjected to, particularly the degree of their confinement. The synergistic effect of shock waves plus blast creates disastrous structural damage and also missile effects of the debris, whereas high temperatures encountered during deflagration practically incinerate everything it comes into contact with. The damages that HEMs can cause can be classified into:

- Formation of highly destructive shock wave and blast pressure in case of high explosives.
- Huge quantities of product gases at high pressures (sometime even up to hundreds of atmospheric pressures) and high temperatures (the flame temperatures of certain propellants can be as high as 3000 K) with enormous heat output when propellants burn.
- Phenomenally high amounts of heat radiation when pyrotechnics burn.

It is, therefore, mandatory that the technical personnel dealing with the HEMs have some fundamental scientific knowledge about their chemical nature, thermal behavior, aspects of

Demystifying Explosives: Concepts in High Energy Materials. http://dx.doi.org/10.1016/B978-0-12-801576-6.00008-2

sensitivity with respect to friction, impact, and static electricity, and problems of compatibility between ingredients that go to make a formulation.

Thermochemical and molecular structural factors and factors like crystal defects, which easily lead to "hot spot" initiation, make quite a few HEMs sensitive to initiation by impact or friction, or heat or discharge of static electricity. This basic knowledge of these aspects is an essential prerequisite for any person who is involved in the synthesis/processing/handling/transportation/storage of HEMs. He/she should be thoroughly aware of these hidden hazards of HEMs.

8.3 Hazard Classification of HEMs

The United Nations have classified different dangerous goods like explosives, toxic chemicals, inflammable chemicals, radioactive materials, etc. under nine categories. Explosives/HEMs are categorized under "**1**". They are further subclassified (1.1−1.6) into six Hazard Divisions (HDs) depending on their sensitivity, as well as the terminal damages they can inflict in case of an accident. Table 8.1 gives a summary of the same. Of these HD 1.1, HD 1.2 and HD 1.3 are highly important.

1. **HD 1.1**: HD 1.1 refers to explosives that undergo mass detonation that creates and propagates shock wave and blast pressure. The destruction is caused mainly by blast and high velocity fragments like shell fragments, boulders, etc. Craters are formed.
2. **HD 1.2**: When there is an accident involving HEMs in cased units (e.g., a rocket motor with nozzle), the major risk is that of propulsion of such a unit and materials of this nature are classified under HD 1.2.

Table 8.1: UN classification of HEMs.

Hazard Division	Effect	Example
HD 1.1	Mass detonation creating shock waves with major blast effects, high blast pressure, & crater.	Initiatories, high explosives
HD 1.2	Projectile and fragmentation hazard	Rocket motor with nozzle, grenades
HD 1.3	Mass fire and radiant heat	Propellants and pyrotechnics
HD 1.4	No significant hazard	Small arms ammunition and caps.
HD 1.5	Very little probability of initiation	No military explosives
HD 1.6	Highly insensitive detonating substance	No military explosives

Table 8.2: Effects of air blast overpressure on human beings.

Probable Effect	Blast Pressure, Psi (kPa)
Ear Drum Rupture	
Threshold	7 (48)
50% Probability	15 (103)
Lung Damage	
Threshold	30–40 (207–276)
Severe	80 (552)
Fatal	
Threshold	100–120 (690–828)
50% Probability	120–180 (828–1242)
100% Probability	200–250 (1380–1725)

3. **HD 1.3**: This includes HEMs like propellants, which undergo mass deflagration (burning). The major risk here is that of **mass fire** and rarely, minor blasts.

The effects of air blast overpressure on human beings have been studied in great detail, and the results are given at Table 8.2.

Out of the three major hazards of HEMs viz. (1) mass detonation, (2) mass fire, (3) thermal radiation, the first two are reversible depending on the conditions like degree of confinement. For example, if we want to burn about 50 kg of gun propellant (like it is done during waste disposal), we should spread it into a thin layer so that the entire surface undergoes only deflagration (burning) safely. If, on the other hand, we make a heap of it, what will start as deflagration in the beginning will transform itself into a detonation because of the confinement. We must understand that confinement refuses to allow the gaseous products to escape, resulting in higher pressures that enormously increase the burning rate of HEMs to such a level that a shock wave is formed. Waste explosive/propellant/pyrotechnic disposal is an extremely hazardous process that has caused many fatal accidents all over the world and, therefore, all precautions/safety norms should be religiously followed during this process.

8.4 The Damages

Many tragic accidents are avoidable by scrupulously following the SOPs/precautionary measures. Before we discuss these procedures/DO's and DONT's, let us remind ourselves that the following are the damages of any major accidents, including HEM–related ones.

1. Personal: Major injuries & Death
2. Property: Buildings/Structures, Facilities, & Materials

3. Morale of Workers
4. Downtime
5. Reputation (of the establishment)

8.5 General Safety Directives

If you are working in the field of HEMs, please pay attention to **EACH and EVERY** point given below.

8.5.1 Assume the Hazard

"Expect the unexpected," particularly while you will work with new materials/compositions.

8.5.2 Never Work Alone!

Work as a group, even if it is a small one.

8.5.3 Start with the Smallest Possible Quantities

Particularly while the compound/composition is expected to be sensitive, e.g., initiatory composition. What should be that "smallest possible quantity" can be decided after thorough discussion with the Safety Division of the establishment.

8.5.4 Safety Shields

Use safety shields—wherever needed.

8.5.5 Fire Hazards: Expect and be Ready

Expect fire hazards and keep your Fire Fighting equipments in readiness.

8.5.6 Ground (Earth) Your Facilities

Grounding/earthing the personnel and equipments is an inescapable requirement when one deals with sensitive HEMs like initiatories and pyrotechnics. In fact, handling propellants (for guns, rockets, etc.) during dry weather also strictly calls for grounding both the working personnel and equipments.

The static electricity discharge pits connected to the equipments should be periodically inspected for their reliability as also the reliability of other static electricity discharge/conducting mats, gloves, and garments.

8.5.7 Wear Protective Garments/Equipments (Including Antistatic Ones)

These include gas masks/goggles/helmets/aprons/safety shoes/antistatic shoes, etc., depending on the type of operation involved.

8.5.8 Practice Relative Humidity Control

When processing/handling explosives, propellants, and pyrotechnics, which are sensitive to static discharge, the Relative Humidity in the process room/laboratory should not be less than 60%. The process rooms should be equipped with humidifiers for this purpose.

8.5.9 Housekeeping

Good housekeeping greatly helps to avoid accidents. Ensure that the labs/process rooms are not cluttered with too many equipments/hardware/materials. Avoid storing incompatible materials together. Ensure before the commencement of operation that the exit pathway is clear.

8.5.10 Know about the Material Hazards

The hazardous nature of materials should be well understood by all the concerned workers/operators. Do thorough literature survey to know such hazards before new processes are tried. (Examples):

1. Chlorates are highly incompatible with carbonaceous matter, ammonium compounds, sulphur, red phosphorus, etc.
2. Water is dangerous with mixtures containing powdered Zr/Ti/Mg/Zn/Al.
3. Very fine ammonium perchlorate/ammonium nitrate can be dangerously sensitive to impact in presence of carbonaceous impurities.

8.5.11 Toxic Hazards

It should be realized that many HEMs and their related chemicals possess not only explosion and fire risks but also toxic hazards. For example, prolonged contact with RDX and trinitrotoluene (TNT) is known to cause skin-related ailments. Isocyanates (like toluene di-isocyanate (TDI) used in composite propellant processing) can cause lungs-related problems like bronchitis. Prolonged ingestion of solvents like benzene might cause cancer while heavy metal ions of barium, and lead might severely impair the functioning of liver and kidneys. Therefore, the following preventive measures have to be taken:

- Compulsory use of personnel protective equipments like gas masks, gloves, aprons, etc., as required

- Periodic workplace monitoring for toxic fumes with reference to the Threshold Limiting Values (TLV)/Short Term Exposure Limit values for the particular chemical
- Effluent treatment if needed

8.5.12 Prepare a Work Plan

- In case of an established process, ensure that Standard Operating Procedure (SOP) has been prepared, taking care of all safety aspects including Man Limit, Explosive Limit, Fire-fighting facilities, Housekeeping, Earthing, etc.
- In case of a new process/synthesis of new HEM, do a thorough literature survey to gauge the hazards involved and then make a step-by-step procedure with precautions to be observed to prevent any runaway reactions/fire/explosion.

8.5.13 Hazard Evaluation

While preparing/processing new explosives/formulations:

1. Start with the smallest quantity.
2. Soon after the initial preparation, evaluate its sensitivity/stability by various tests like Impact sensitivity, Friction sensitivity, Spark sensitivity, differential thermal analysis (DTA), Vacuum stability, etc.
3. In case of new mixtures, first evaluate the compatibility between various ingredients using techniques like DTA.

The results of these tests will adequately caution you before you do further processing/ scaling up.

8.5.14 Storage/Transport

During storage/transport of explosives, due care should be taken to observe the statutory explosive regulations very meticulously. While planning an explosives process building or magazine, various safety distances like Storage Inside Quantity Distance (SIQD), Process Inside Quantity Distance (PIQD), and Outside Quantity Distance (OQD) should be strictly followed apart from the type of protection necessary like the requirement of a particular type of traverse, blast wall, etc. Both during storage and transport of explosives, care should be taken to ensure that:

- only the approved type of package and transport like explosive van should be used.
- no incompatible groups of explosives are transported together.

Note: Extensive studies and trials have been carried out to decide upon the Quantity–Distance relation in the field of explosives. For instance, when one wants to construct a magazine, an Explosive Storage House (ESH) for storing 2 tons of RDX (categorized under HD 1.1), what should be the minimum safety distance (D) from the ESH to another similar ESH as well as to a residential colony? Naturally, the value of D cannot be same for both, and in the latter case, it should be far greater than the first one. We are guided by an empirical formula given below to determine the minimum safety distance, D required in such as case:

$$D = K \times Q^{1/3} \tag{8.1}$$

where D = minimum distance required between the ESH (sometimes referred as Potential Explosion Site) and the building/installation/infrastructure under consideration (measured in meters).

Q = Net Explosive Quantity in kilogram at ESH
K = Protection level, the value of which depends on what you want to protect.

Figure 8.1 explains this concept.

In this example (where NEQ is 2000 kg), the values of K for another ESH (magazine) and residential colony are 2.4 and 22.2, respectively. Accordingly:

$$(\text{SIQD}) : D = 2.4(2000)^{1/3} \sim 31\,\text{m} \quad (\text{minimum})$$
$$(\text{OQD}) : D = 22.2(2000)^{1/3} \sim 280\,\text{m} \quad (\text{minimum})$$

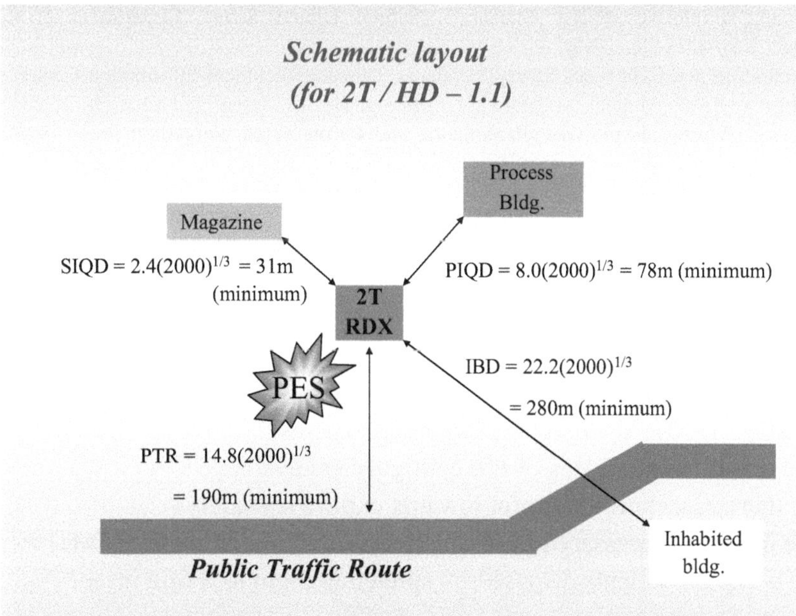

Figure 8.1
Typical Representation of Quantity Distance Relation.

It means that there is a nine-fold increase in the minimum (safety) distance when we compare a residential colony with another ESH.

8.5.15 Waste Disposal

Although it may appear innocuous and routine, waste disposal of explosives, propellants, and pyrotechnics is probably one of the most hazardous operations in the field of explosives. As already mentioned earlier, many fatal accidents have been well-reported during the waste disposal of explosives and ammunitions. Their disposal should be well planned and carried out strictly as per the laid down norms available in the literature.

8.6 Conclusion

As it is normally described about fire and electricity, explosives are our "best friend but also our worst enemy." Remember that when we talk about safety:

- Ignorance cannot be excused
- Negligence cannot be tolerated
- Overconfidence cannot be pardoned

Suggested Reading

[1] R.M. Downey, Explosives Safety Standards: Safety, United States, Department of the Air Force, Headquarters US Air Force, 1992.
[2] DoD, Ammunition and Explosives Safety Standards, Defense Technical Information Center, 1978.
[3] DOE Explosives Safety Manual, Manual HS—Office of Health, Safety and Security, January 09, 2006.
[4] A. Bailey, S.G. Murray, Explosives, Propellants, and Pyrotechnics, Pergamon Press, Oxford, New York, 1988.
[5] Service Textbook of Explosives, Ministry of Defence, Publication, UK, 1972.
[6] P.W. Cooper, Explosives Engineering, VCH, Publishers Inc, USA, 1996.
[7] J. Akhavan, The Chemistry of Explosives, third ed., Royal Society of Chemistry, 2011.

Questions

1. What are the different hazard classifications of HEMs?
2. What is SOP? How it is important for new processes?
3. What are the different classes of fire extinguishers available?
4. How are thermal techniques useful towards explosives safety?
5. What are the steps necessary to prevent electrostatic initiation of HEMs?
6. Why are waste propellants disposed by spreading them as a thin layer?

CHAPTER 9

HEMs: Concerns of Security

9.1 HEMs: Concerns of Security

Palpably, terrorism is the number one menace and threat to global peace today. The most common tools that the terrorists use today are high explosives, although the world should be ready to prevent and combat terrorism based on more disastrous tools like nuclear, biological, and chemical weapons. The very survival of humanity today depends on the human will, technological advancement, and judicious strategies in this direction.

We are witnesses to the use of high explosives in terrorist attacks in versatile ways right from the crude lumps containing simple mixture of ammonium nitrate and nails (intended to be high-velocity projectiles on the initiation of AN) and a detonator to sophisticated, remote-operated explosive devices. When the terrorists fail to get stolen ammunition or relatively costly and strategic explosives like RDX, the option of easily accessible civil explosives (mostly AN-based and some times NG/dynamite-based) is always open to them. The explosive devices used in unconventional warfare by terrorists are referred as **Improvised Explosive Devices (IEDs)** and they can take any form like letter bombs, pipe bombs, or explosive devices kept in a radio transistor/suitcases/lunch boxes/toys, etc. Some of the commonly used explosives in IEDs are given in Table 9.1.

The use of innocuous materials as explosives for terrorist activity is a cause of worry. Recent approaches to use CHO materials (free from nitro and nitrato groups to escape detection) are an alarming trend. For example, it is reported that triacetone triperoxide (TATP) was about to be used in the terrorist attempt foiled a few years back in London. It was intended to blast the aircrafts in midair. It can be obtained in crude state from polish remover. Hexamethylene triperoxide diamine (HMTD) is another compound of this class, which was captured from Algerian terrorists entering into the United States from Canada.

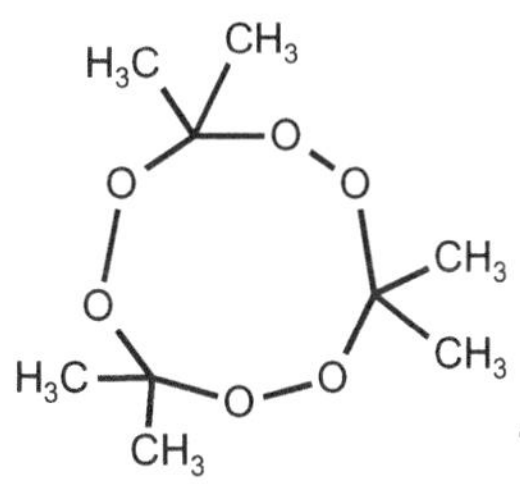

TATP

Demystifying Explosives: Concepts in High Energy Materials. http://dx.doi.org/10.1016/B978-0-12-801576-6.00009-4

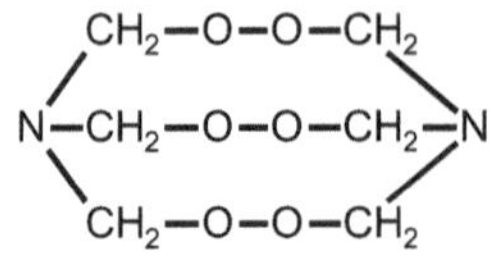

HMTD

Table 9.1: Some improvised explosive devices (IED) compositions.

Conventional/Military Explosives used in IEDs	Commercial Explosives used in IEDs
RDX-based IEDs	**Ammonium nitrate–based IEDs**
SEMTEX (RDX, styrene-butadiene copolymer & additives (plastic explosive used in 1988 Pan Am aircraft blast))	*Red diamond*: Ammonium nitrate, Sodium nitrate, Nitroglycerine, & additives
SEMTEX-H (RDX, PETN, styrene-butadiene copolymer, motor oil, & additives	*ANFO*: Ammonium nitrate & fuel oil
C-2: RDX, TNT, DNT,[a] MNT, & NC	*Prillex*: Ammonium nitrate & diesel oil
C-3: RDX, TNT, DNT, Tetryl, & NC	*Sigmagel Titagel*: Ammonium nitrate, Sodium nitrate, & Calcium nitrate
C-4: RDX, Polyisobutylene, & Fuel oil)	*Lovex*: Ammonium nitrate, mono-methyl ammonium nitrate, & gelling agent
	Emulsion explosives
	Nipak: Ammonium nitrate, Sodium nitrate, polyurethane, & additives
TNT based IEDs	**Miscellaneous**
Cyclotol: RDX & TNT	Petrogel: Nitroglycerin, Ethylene Glycol Dinitrate, Nitrocellulose, sodium nitrate, & additives
Tetryol: TNT & Tetryl	
PETN based IEDs	
Detsheet: PETN & Plasticizer	*Dynamite*: NG + Keiselgur
Pentolite: PETN & TNT	Slurry and water gel explosives

[a]DNT, Dinitro toluene.

9.2 Detection of Explosives

The detection of hidden explosives and prevention of a disaster is one of the major technological challenges today. Although a huge amount of work has been and is being done in this direction, different devices designed and manufactured for this purpose have their own advantages and disadvantages. One of the earliest methods adopted was to make it a statutory obligation on the part of an explosives manufacturer to add certain chemicals in a small percentage to the explosives at the time of processing. The said chemicals (called taggants) have a low vapor pressure, but their vapors are easily detectable by devices such as Electron Capture Detector (ECD). However, if the IED is thoroughly sealed, hardly allowing any vapor of that chemical to effuse out, this method will be of no use. Some such taggant chemicals are given below:

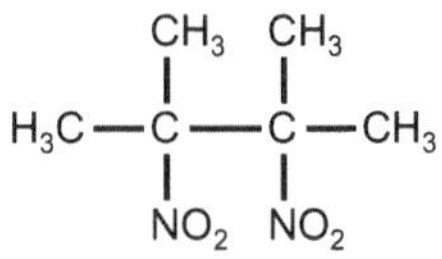

2,3-Dimethyl-2,3-dinitrobutane

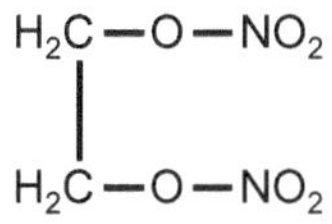

Ethylene glycol dinitrate

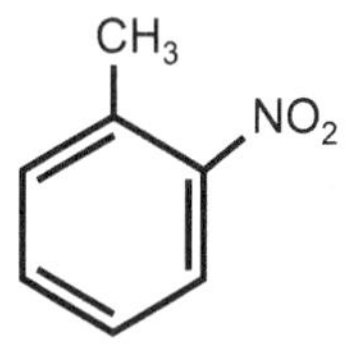

Ortho mononitro toluene

para mononitro toluene

(*Note*: Most of the explosives themselves have very low vapor pressures. For example, the vapor pressures of RDX and PETN (in mm of Hg at 25 °C) are 8.0×10^{-8} and 7.0×10^{-9}, respectively. In case these explosives are embedded in a polymeric matrix as a plastic explosive, the vapor emission will go down further drastically).

In the detection of explosives, sniffer dogs have application since as long as mobile detectors have been around. It is reported that they have about 90% reliability. However, major problems are their deployment in public places, need for continuous training, and proper handlers. Law enforcing agencies are increasingly dependent on conventional X-ray detectors at entry points like airports, seaports, and other important public places. Although such heavy X-ray detectors have been doing a good job in scanning the baggages to detect any explosive devices, their immobility limits their use in detecting hidden explosives elsewhere. At times, it may be required to detect hidden explosive at a stand-off distance in view point of safety. For such purposes, stand-off detectors are designed to detect explosives at a distance of 10 m or more. Different devices have been and are still being developed for the purpose of detection of explosives, and each of them is based on a specific principle such as electron capture (Electron capture detector (ECD)), chemiluminescence (CL Detector), ion mobility (Ion mobility spectrometer (IMS)), diamagnetism of materials, fast neutron activation, etc., and a few of them are described below:

9.2.1 Electron Capture Detector

Principle: It records changes in current due to absorption of electrons by certain electron-absorbing groups (e.g., NO_2) present in explosives molecules. The ECD is used for detecting electron-absorbing components of high electronegativity such as halogenated compounds in the output stream of a gas chromatograph.

Advantages	Disadvantages
Highly selective	Only usable for a few constituents
High sensitivity (<1 pg detection limit)	Radioactive detector is used
Nondestructive	Smaller linear range & response factors vary considerably

9.2.2 Ion Mobility Spectrometer

Principle: It records the mobility of the explosive molecular ions that is characteristic of an explosive. IMS is a spectrometry technique capable of detecting very low concentrations of chemicals based upon the differential migration of gas phase ions through a homogeneous electric field.

Advantages	Disadvantages
Detects the presence or absence of an energetic material in seconds	Low resolution
(Can detect quantities from 0.1 to 10 nanograms)	Susceptible to atmospheric changes

9.2.3 Thermoredox Detector

Principle: It records the electrochemical reduction of $-NO_2$ group present in the explosives. This technology is based on decomposition of explosive substance followed by the reduction of the NO_2 groups.

Advantages	Disadvantages
This technique does not require a carrier gas other than ambient air	Sensitivity is fairly low
System is portable, lightweight, and powered by rechargeable batteries	Harmless nitro compounds often create false alarms
Low consumable cost, requires very limited operator training, and user friendly	Suitable only for compounds with high vapor pressure

9.2.4 Field Ion Spectrometer

Principle: The principle is based on filtering ion species according to the functional dependence of their mobilities with electric field strength.

Field ion spectrometer, also known as transverse field compensation IMS, is a new technique for trace gas analysis that can be applied to the detection of explosives and narcotics. It eliminates the gating electrodes needed in conventional IMS to pulse ions into the spectrometer; instead, ions are injected into the spectrometer and reach the detector continuously, resulting in improved sensitivity. The technique enables analyses that are difficult with conventional, constant field-strength IMS.

9.2.5 Diamagnetism-Based Magnetic Field Detector

Principle: The detection is based on the principle that every material has a characteristic magnetic property and can be detected accordingly.

Magnetometers have a wide range of potential applications, and where there is an electrical current, there is a magnetic field. Measurements of magnetic fields can reveal information about the electrical activity, the chemical identity of a spinning atom, or simply the presence or absence of metal. This consists of a laser, a cell containing vaporized metal atoms, and a light detector. When the metal atoms are illuminated by the laser, they align such that they don't absorb any of the light. The presence of even a very weak magnetic field, however, disrupts their alignment, and they absorb some of the light. This change is recorded by the detector.

With small size and sensitivity, the new sensors promise to improve detection of bombs and could be incorporated into future magnetic resonance imaging (MRI) scanners. It is small and cheap, and uses very little power. For the detection of IED or unexploded ordnance in minefields, the small size and low power consumption of the sensors could make a big difference. The sensors could be grouped in arrays, making it possible to gain more data in a given amount of time.

9.2.6 Nuclear Quadrupole Resonance Detector

Nuclear Quadrupole Resonance (NQR) is a sensor technology related to nuclear magnetic resonance (NMR). Any nucleus with more than one unpaired nuclear particle (protons or neutrons) will have a charge distribution that results in an electric quadrupole moment. NQR measures a signature unique to the explosive contained in the hidden objects, thus providing a means of efficiently detecting land mines.

NQR can detect even small quantities of explosives. NQR signature is independent of the shape of the explosive. The signature emanates directly from the condensed phase, and NQR does not have the shortcomings that plague vapor-phase chemical detectors. It provides the chemical specificity of NMR and the volume capacity of MRI without the need for expensive and cumbersome DC magnets.

9.2.7 Micro Electro Mechanical Systems

Micro Electro Mechanical Systems (MEMs) is a recent technology and it consists of integrated mechanical elements, sensors, actuators, and electronics on a silicon substrate using a process technology called microfabrication. The sensors gather information by measuring mechanical, thermal, biological, chemical, magnetic, and optical signals from the environment. The microelectronic integrated circuits (ICs) act as the decision-making piece of the system by processing the information given by the sensors. Finally, the actuators help the system respond by moving, pumping, filtering, or somehow controlling the surrounding environment to achieve its purpose. Research and development efforts are in progress to develop a viable and general purpose explosives detection system based on MEMs.

A number of explosive vapor detection devices based on other spectroscopic techniques like photoluminescence, Resonance enhanced multi-photo ionization, Cavity ring down spectroscopy, Laser induced breakdown spectroscopy, Raman Scattering and Laser imaging detection and ranging, etc., are emerging on the scene. The technologies receiving major attention are described below.

1. Biosensors
2. Surface Acoustic Wave
3. Micro cantilever–based mine detection system
4. Amplifying fluorescent polymers
5. Detector-based on diamagnetism

During recent times, the miniaturization of analytical instruments has resulted in the availability of UV–VIS, Near infra red (IR), fluorescence, Raman spectrophotometers for field applications. The literature reports also indicate that the miniaturization of mass spectrometers has also been mastered and they may become available in near future for field analysis.

The trace explosive detectors require the operator to approach the IED to close distances, of the order of a few centimeters, unless some robot or unmanned ground/aerial vehicle is employed. The detection of an IED in large area like residential area or ground/stadium becomes a laborious and time-consuming task.

In view of the unabated use of different explosives by terrorists with varying degrees of innovation and sophistication, huge sums of money are being spent toward the development of detectors with better accuracy and reliability, portability, very low probability of setting false alarms, and safety features.

Suggested Reading

[1] J. Yinon, Forensic and Environmental Detection of Explosives, John Wiley & Sons, Inc, 1999.
[2] M. Marshall, J.C. Oxley, Aspects of Explosives Detection, first ed., Elsevier Science, 2011.
[3] J. Yinon, Counterterrorist Detection Techniques of Explosives, Elsevier, 2007.
[4] J. Gardner, Y.J. Jehuda, Electronic noses and sensors for the detection of explosives, in: Proceedings of the NATO Advanced Research Workshop, Held in Warwick, Coventry, U.K, 2003.
[5] J. Gardner, Y. Jehuda, Electronic Noses and Sensors for the Detection of Explosives–NATO Science Series II, 2004. New York.
[6] H. Schubert, A. Kuznetsov, Detection of explosives and landmines methods and field experiences methods and field experience, in: Proceedings of the NATO Advanced Research Workshop, Petersburg, Russia, 2001.
[7] H. Schubert, A. Kuznetsov, Detection and disposal of improvised explosives, in: Proceedings of the NATO Advanced Research Workshop on Detection and Disposal of Improvised Explosives St. Petersburg, Russia, 2005.
[8] H. Schubert, A. Kuznetsov, Detection of liquid explosives and flammable agents in connection with terrorism, in: Proceedings of the NATO Advanced Research Workshop on Detection of Liquid Explosives and Flammable Agents in Connection with Terrorism, NATO Science for Peace and Security Series B, Petersburg, Russia, 2007.

Questions

1. What is meant by IEDs?
2. Why are IEDs difficult to be detected?
3. What are taggants? Name any two taggants used for military explosives.
4. How does an ECD work?
5. What are MEMs? How are they fabricated?
6. Are there any common methods of detection of explosives?

CHAPTER 10

HEMs: Characterization and Evaluation

10.1 Introduction

In any research area concerned with new chemical compounds, characterization and evaluation of the synthesized compounds, including intermediates, are of vital importance. High-energy materials (HEMs) cannot be an exception. Characterization is essentially an identification process whereas evaluation refers to the measurement of certain special characteristics of the synthesized compounds. For instance, a newly synthesized energetic compound is "characterized" by a systematic process involving chromatography (to ensure the purity of the compound), spectroscopy, and any such method to make sure of its molecular structure so that the chemist knows what exact compound he has synthesized. On the other hand, the chemist would "evaluate" the new compound for certain specific performance parameters or performance potentials. For example, a newly synthesized explosive molecular compound may be evaluated for calorimetric value (thermochemical potentials), velocity of detonation (VOD; detonation potential), or friction/impact sensitivity figures (mechanical sensitivity potential).

With the advent of highly sophisticated instrumental analytical techniques, the characterization and evaluation techniques related to HEMs have come a long way over the years. Chromatography, spectroscopy, and thermal analysis techniques are the mainstay for characterization and evaluation of HEMs. It might be interesting to note that certain techniques that are as old as a century or several decades are still being followed today when it comes to the evaluation of certain characteristics of HEMs. The vacuum stability test for explosives and certain propellants, friction and impact sensitivity tests for almost all HEMs, and shock sensitivity tests for explosives may look a little archaic or even outdated; however, all of these tests are time-tested, highly reliable, and totally indispensable. Apart from these, there may be highly specific tests for a particular explosive. For instance, for nitrocellulose (NC), the Bergmann and Junk test chemically measures the amount of oxides of nitrogen evolved on heating a gram weight of NC for a specific period at a specific temperature. The amount of oxides of nitrogen evolved (evaluated titrimetrically) indicates the extent of instability of NC. This chapter does not include such tests, but it gives a general approach to the characterization and evaluation of HEMs.

Demystifying Explosives: Concepts in High Energy Materials. http://dx.doi.org/10.1016/B978-0-12-801576-6.00010-0

10.2 Chromatographic Techniques

Chromatographic techniques are a group of analytical techniques used for the separation of components from a mixture using differences in their distribution between two phases (stationary and mobile phase). Many chromatography techniques are available today for characterization purposes (e.g., thin layer chromatography (TLC), gas chromatography (GC), high-performance liquid chromatography (HPLC) etc.). These techniques are used for identification, separation, characterization, and quantification. Let us discuss only the techniques that are frequently used for the characterization of energetic materials.

10.2.1 Thin Layer Chromatography

TLC is a quick, simple, ready-to-use, inexpensive tool typically used in laboratory synthesis practices. It gives the quick answer to the synthesis chemist whether he/she is going in the correct direction or not. This is usually not considered a characterization technique, but it gives an idea about the number of components in a reaction mixture. TLC is used to confirm the presence of unknown substances in comparison to the standard known substance using the relative front (R_f), which is the ratio of the distance traveled by the solvent front to the distance traveled by the substance under examination. TLC is also used to monitor the course of the reaction and gives an idea about the conversion of reactant to product. TLC experiments are generally performed with reactant, product, and possible byproduct in a single run.

A TLC plate is a thin sheet of solid adsorbent (usually silica or alumina) spotted with known and unknown substances. This plate is eluted with a proper solvent (often binary mixtures based on the polarity) liquid. Once the solvent reaches the top of the plate, the plate is removed from the developing chamber, dried, and the separated components of the mixture are visualized in an ultraviolet (UV) lamp. Identical compounds possess similar R_f, and dissimilar compounds deviate up or down.

10.2.2 Gas Chromatography

In GC, the mobile phase is an inert gas and the stationary phase is a liquid or solid. Thus, in GC, separation of components in a chemical mixture is achieved on the basis of differences in the partition coefficient of solutes in the gas phase and stationary phase. In GC, the time gap between the injections of the sample (zero time) and the peak maximum of substance is called its retention time (RT). It is the characteristic property of the compound, and it varies from compound to compound. The greater the affinity of the compound for the stationary phase, the more the compound will be retained by the chromatographic column and will be eluted later than the one having less affinity for the stationary phase. The major limitation of GC is that the compound analyzed should possess reasonably high vapor pressure. Low-melting explosives (e.g., trinitrotoluene and 2,4-dinitroanisole (DNAN)) and compounds with high vapor pressure (volatile substance and liquids) mostly can be analyzed by GC. The main advantage of GC is that the analysis is faster and accurate.

10.2.3 High Performance Liquid Chromatography

In contrast to GC, HPLC uses a liquid as the mobile phase, and liquid phase (coated on inert solid support), solid adsorbent (e.g., silica or alumina), or ion-exchange resin is used as the stationary phase. The separation of constituents in HPLC is based on the interaction of the individual components, and the stationary phase and components are retained to a different extent, which causes the separation. For example, those samples that have stronger interactions with the mobile phase than with the stationary phase will elute from the column faster and thus have a shorter RT. Likewise, those who do have strong affinity to the stationary phase will stay in the column for longer duration.

HPLC is superior to any other liquid chromatographic techniques in terms of separation efficiency. The analysis of mixtures can be done faster with HPLC because of the increased flow rate using high-pressure pumps.

Reverse-phase HPLC is the method of choice for the detection and quantification of explosive molecules. In the analysis of nitro compounds, a UV detector is mainly used. In this technique, several components can be identified and quantified in a short time.

Many explosives and their intermediates have been analyzed by HPLC. An example of an HPLC chromatogram is shown in Figure 10.1. It can be seen that the order of the elution

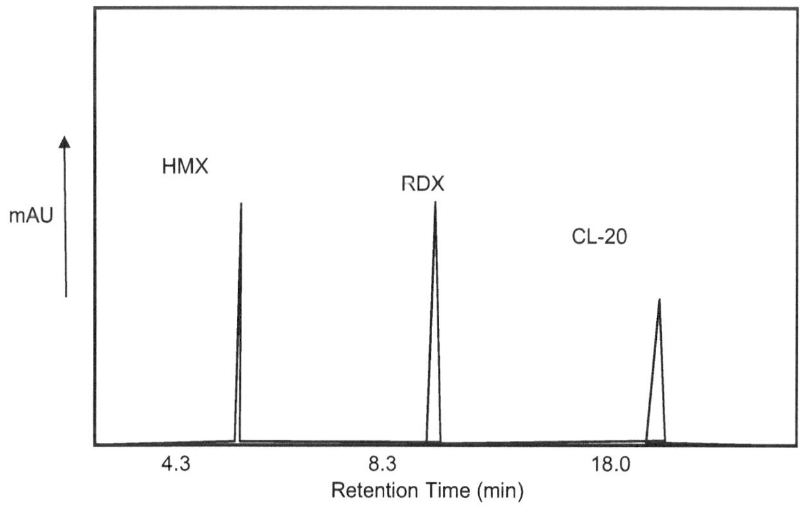

Figure 10.1
HPLC Chromatogram of Nitramine Explosives.

is (1) HMX (4.3 min), (2) RDX (8.3 min), and CL-20 (18 min) with the following instrumental parameters: mobile phase, water/methanol (60:40); flow rate, 1.2 mL/min; injection volume, 10 μL; and column, C-18.

10.3 Spectroscopic Techniques

Spectroscopy is a well-known analytical technique for the identification of functional groups in chemical substances. In spectroscopy, a certain portion of the electromagnetic spectrum (UV, visible (VIS), or infrared (IR)) interacts with matter and the resultant spectrum is interpreted to diagnose the molecular structure of the chemical substance.

10.3.1 UV/VIS Spectroscopy

UV radiation (wavelength varying from 200 to 400 nm) is the part of electromagnetic radiation that can promote the electrons of a molecule from their ground state to an excited state. The VIS portion of the spectrum lies between 400 and 800 nm. UV and VIS spectrometers typically are available together. Both regions correspond to energy level characteristics of excitation of π and nonbonding electrons and are most often associated with molecules containing conjugated double bonds. This spectroscopic method provides only limited information to the chemists. Fortunately, most of the explosives molecules possess groups containing π electrons and n electrons and are UV active. For example, $-NO_2$ groups present in explosive CL-20 appear as a broad peak at 230 nm in the UV spectrum. Some more examples are given in Table 10.1.

10.3.2 IR Spectroscopy

The IR region of the electromagnetic spectrum (4000 cm^{-1} to 400 cm^{-1}), which corresponds to changes in vibrational energies within molecules, is very helpful for the identification of functional groups in chemical characterization. Not all possible vibrations within a molecule will result in an absorption band in the IR region. For a molecule to be

Table 10.1: Some UV Active Explosives.

Compound	λ_{max} (nm)	Compound	λ_{max} (nm)
NB	269	1,3-DNB	242
1,3,5-TNB	227	2,4,6-TNT	232
Picric acid	378	Picramide	333
RDX	213	HMX	228
CL-20	230	Nirtroguanidine	265

λ_{max}, wavelength at maximum intensity; NB: nitrobenzene; DNB: dinitrobenzene; TNB: Trinitrobenzene

Table 10.2: IR Absorption Frequencies.

Groups	IR Peak (cm^{-1})	Groups	IR Peak (cm^{-1})
C–H	2850–3000	$-NO_2$	(1) 1510–1560 (2) 1330–1370
O–H	3000–3400	C≡N	2220–2260
N–H	3100–3450	$-N_3$	2200
C=O		$-NO_3$	1350–1380
Aldehydic	1680–1740		
Ketonic	1665–1725		

Note: cm^{-1} is the unit of wave number, which is the reciprocal of wavelength in cm.

IR active, the vibration must result in a change of dipole moment during the vibration. The IR absorption frequencies of some of the typical functional groups found in most explosives are given in Table 10.2.

Explosive samples are generally analyzed by mixing a small quantity of sample and a mineral oil nujol to give a paste, which is then applied between two sodium or potassium chloride plates. The plate is then fitted into the IR instrument and analyzed. Another method of analysis is by mechanically pressing the finely ground sample and pure potassium bromide (KBr) into a transparent disc in a die under pressure. Later, the KBr disk containing the sample is placed in a sample holder ready for scanning in an IR machine.

Nowadays, Fourier transform IR (FTIR) spectrophotometers are used for the analysis. Analysis is faster in FTIR, and it takes a few seconds to record the spectrum. Another advantage is that a very small quantity of substance is sufficient to record a reasonably good spectrum.

10.3.3 Nuclear Magnetic Resonance Spectroscopy

In a molecule, every spinning proton acts as a tiny magnet. Therefore, molecules containing atoms such as H^1 and F^{13}, which have an intrinsic magnetic moment, can interact with an external magnetic field giving rise to nuclear spin energy levels. When molecules containing one or more hydrogen atoms are placed in a magnetic field, the magnetic moment of the proton gets aligned. For a proton, the quantum theory permits only two orientations that differ in energy, the energy separation being proportional to the strength of the magnetic field. In typical experiments, the energy gap is so adjusted that 60 megacycles (6×10^7 cP) of electromagnetic radiation (corresponds to radiofrequency (RF) of the electromagnetic spectrum) is able to cause transition between the energy levels.

The utility of the magnetic resonance method arises from the fact that the local molecular environment of a hydrogen atom in a molecule slightly perturbs the energy gap, thus

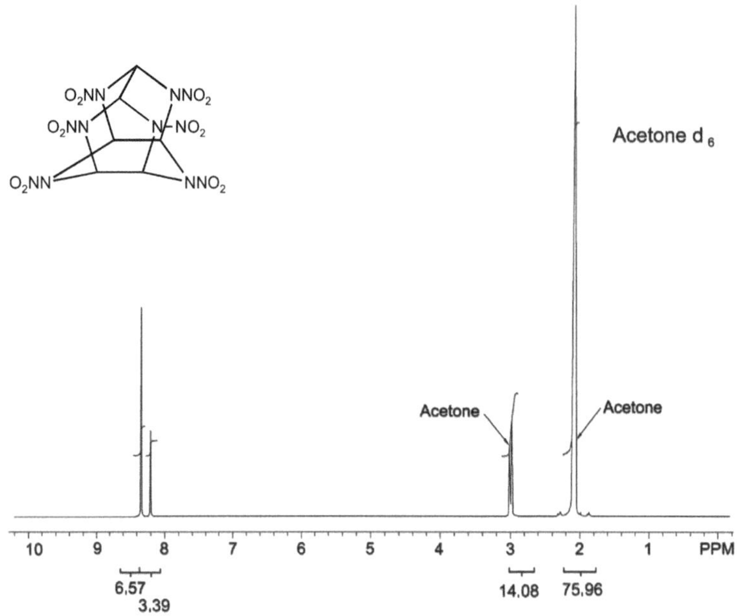

Figure 10.2
NMR Spectrum of CL-20.

modifying the frequency of the absorbed radiation. For example, in a molecule of ethanol (CH_3-CH_2-OH), the H atoms present in CH_3, CH_2, and OH have different molecular environments and they produce their own characteristic shifts in the RF absorption called "chemical shifts." Measurement of chemical shifts yields accurate information about the total number of H atoms as well as the type of H atom (e.g., C–H, O–H, N–H, etc.), helping the chemist to elucidate the molecular structure of a given compound.

Nuclear magnetic resonance (NMR) spectroscopy is an important tool for structural elucidation during the synthesis and analysis of HEMs. A typical such NMR spectrum of the explosive CL-20 is given in Figure 10.2.

10.4 Thermal Evaluation of Energetic Materials

The thermal evaluation of energetic materials is an important area in assessing the performance and suitability of the material for various applications. In thermal evaluation techniques, a small quantity of energetic material under investigation is subjected to programmed heating, and the response from the sample is recorded with respect to temperature.

Thermal evaluation techniques enable one to obtain better insights on the following aspects: thermal stability, shelf life, compatibility, safety aspects, transition temperature, heat capacity, melting temperature, crystallization kinetics, hazard evaluation, aging and thermal history effects, quality control, dehydration, dehydration kinetics, heat of transition, phase transition, glass transition, etc. The important techniques generally used are as follows:

1. Differential thermal analysis (DTA)
2. Differential scanning calorimetry (DSC)
3. Thermogravimetric analysis (TGA)
4. Simultaneous thermal analysis (STA)

10.4.1 Differential Thermal Analysis

DTA is the simplest form of a thermal analysis technique. The principle of DTA is based on subjecting the sample (e.g., ammonium perchlorate [NH_4ClO_4, AP]) and an inert reference material (mostly aluminum oxide) to a simultaneous temperature program and recording the differential temperature (i.e., difference between the temperature of the sample and that of the reference, i.e., ΔT) with respect to the temperature. A few milligrams each of the sample and reference are taken in separate platinum cups connected to Pt or Pt/Rh temperature sensors. The entire setup is kept in a heating furnace, and the sample and reference materials are heated at a specified heating rate (e.g., at 10 °C/min). The difference in temperatures of sample and reference (ΔT) is recorded on the ordinate and the temperature (or time, because the rate of heating is a constant) is recorded on the abscissa.

As long as no reaction (chemical reaction and physical changes such as melting, phase changes, etc., which involve heat changes) takes place in the sample, ΔT is zero. If the sample undergoes an endothermic change (e.g., melting), then its temperature goes lower than the temperature of the reference and ΔT is recorded as negative. Conversely, if the sample undergoes an exothermic reaction (e.g., oxidation), then its temperature will be higher than that of the inert reference material and ΔT is recorded as positive. A DTA thermogram (Figure 10.3) is given for AP and the possible explanations are given (based on other evidences).

a. ΔT: −ve; endothermic (240°C) Phase change of AP (orthorhombic to cubic)
b. ΔT: +ve; exothermic (290°C) First stage oxidation of AP
c. ΔT: +ve; exothermic (360°C) Complete oxidation of AP resulting in its deflagration

10.4.2 Differential Scanning Calorimetry

DSC is the most important thermal evaluation technique. In DSC, a known quantity of sample and a reference are subjected to programmed heating and the difference in energy inputs into the sample and reference is measured as a function of temperature. The DSC

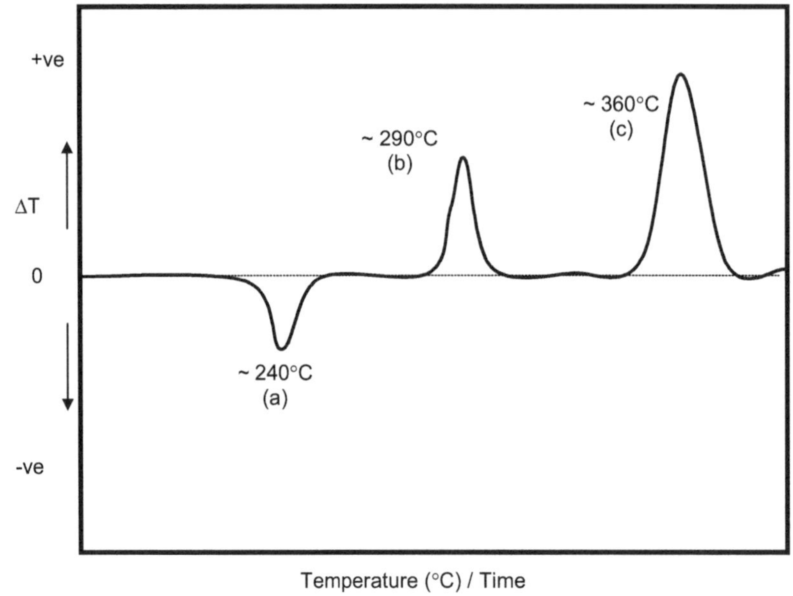

Figure 10.3
DTA Thermogram of Ammonium Perchlorate (AP).

technique is more quantitative, accurate, and faster than the DTA method. In DSC, the temperature of the sample and reference is not allowed to vary. For instance, during simultaneous temperature programming, if the sample undergoes an exothermic reaction, (thereby becoming hotter than the reference), then the reference is given the heat output (which can be accurately measured) such that it also attains the temperature of the sample; conversely, in an endothermic reaction of the sample, heat output is given to the sample. In a DSC thermogram, the heat output is plotted against the temperature. Figure 10.4 shows the DSC thermograms of ammonium dinitramide (ADN) and CL-20. A DSC thermogram of ADN shows two endothermic events and one exothermic event. The first endothermic peak is due to the melting of ADN (92 °C), and the second peak corresponds to the exothermic decomposition of ADN (150–250 °C with a T_{max} of 184 °C) to various decomposition products (mainly ammonium nitrate). Ammonium nitrate formed during the decomposition sublimes (endothermic peak at 264 °C) in the third step.

CL-20 exists in various polymorphs such as α, β, δ, ε, γ, etc., but the ε form of CL-20 is more stable at ambient temperature than other polymorphs. The DSC thermogram of the ε form of CL-20 shows one small endothermic and a significant exothermic peak (Figure 10.4(b)). The ε form of CL-20 absorbs heat energy and is transformed into the γ form of CL-20 at 165 °C. This event appears as an endothermic peak. During further heating, CL-20 exothermically decomposes (onset at 220 °C with a T_{max} of 252 °C) into various decomposition products and releases high heat output.

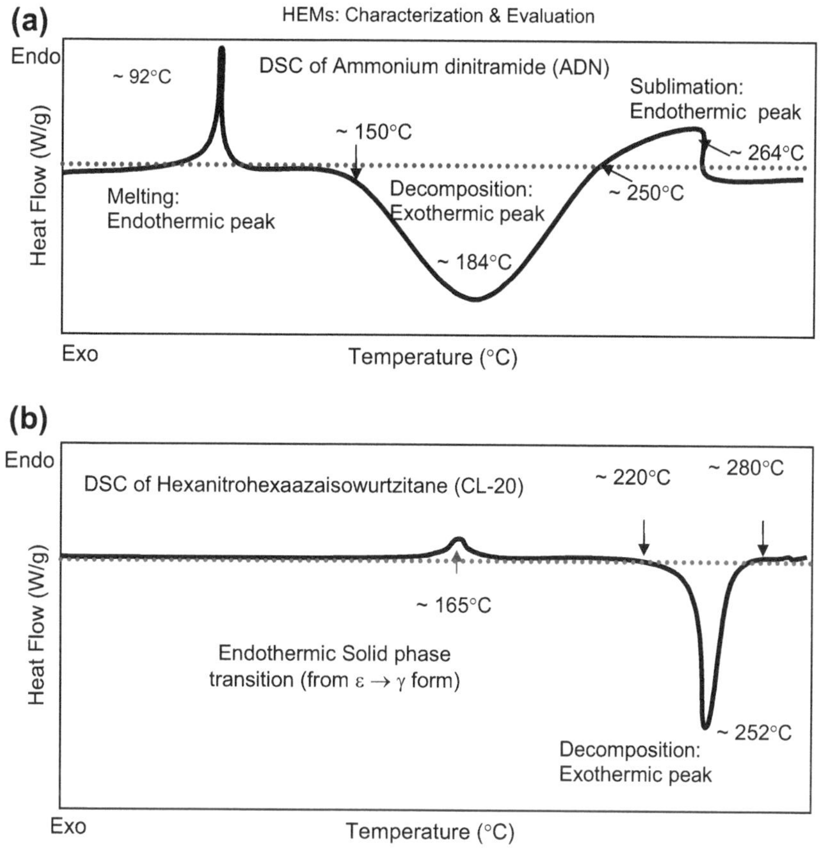

Figure 10.4
(a) DSC Thermogram of ADN and (b) DSC Thermogram of CL-20.

DSC study helps an HEM chemist to evaluate the compatibility of ingredients in an explosive formulation. The compatibility assessment is based on the principle that the addition of an ingredient (the compatibility of which with the explosive is being assessed) should not bring down the decomposition temperature of the virgin explosive by more than 5 °C. To conduct the assessment, the DSC thermograms are taken for the main explosive and for the explosive to which the ingredient is added to the extent required by the formulation. Let us propose to add either plasticizer A or plasticizer B as 5% of the formulation to NC (i.e., 95% NC + 5% plasticizer). The DSC thermograms taken (Figure 10.5) show that plasticizer A is compatible whereas plasticizer B is not.

10.4.3 Thermogravimetric Analysis

In the TGA thermal evaluation technique, a known quantity of the sample is subjected to programmed heating, and the weight loss pattern of the sample is measured as a function

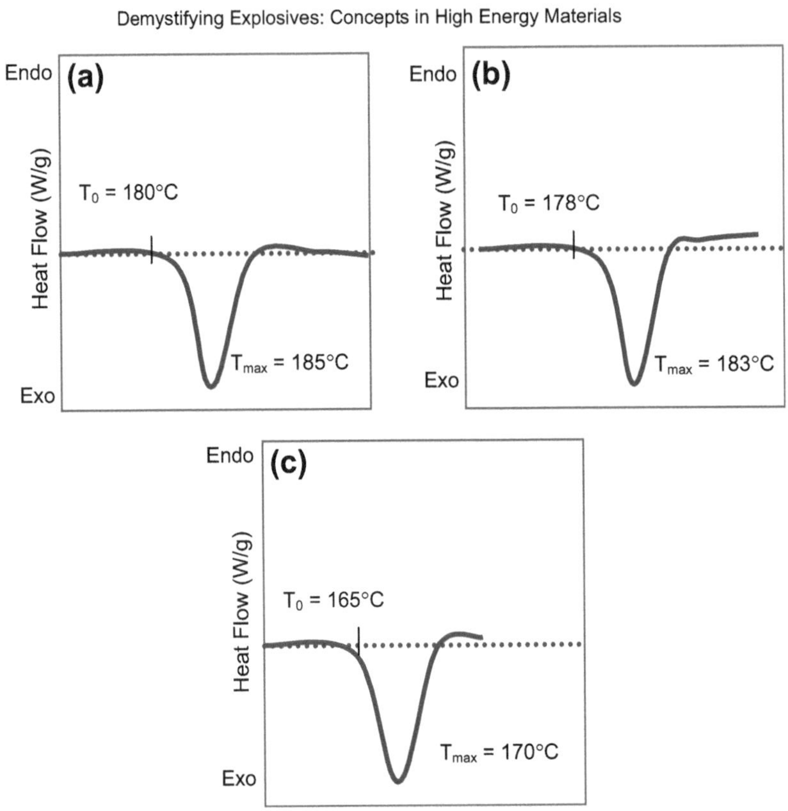

Figure 10.5
(a) DSC of NC, (b) DSC of NC + Plasticizer A (95:5), and (c) DSC of NC + Plasticizer B (95:5).

of temperature or time. The initial weight of the sample is considered as 100%, and the loss in weight is recorded as a percentage. As explosives are heated, they lose their weight through various processes such as dehydration, evaporation, or decomposition. It is interesting to note that some materials can gain weight by reacting with the atmosphere in the testing environment. The weight loss data are usually plotted on the *y*-axis and temperature/time on the *x*-axis. Thermogravimetric evaluation of explosives offers information about the thermal stability of explosives.

TGA of dinitroanisole (DNAN) and *N*-methyl-2,4,5-trinitroimidazole (MTNI) is depicted in Figure 10.6. DNAN loses its mass in a single stage in the temperature region of 97–225 °C with a mass loss of 97%. MTNI loses its mass in two stages. In the first stage, 87.5% mass is lost in the temperature region of 105–235 °C, and in the second step 10.5% is lost in the temperature region of 235–320 °C. Apart from obtaining an idea about the thermal stability of explosives, TGA may help in obtaining insight into the

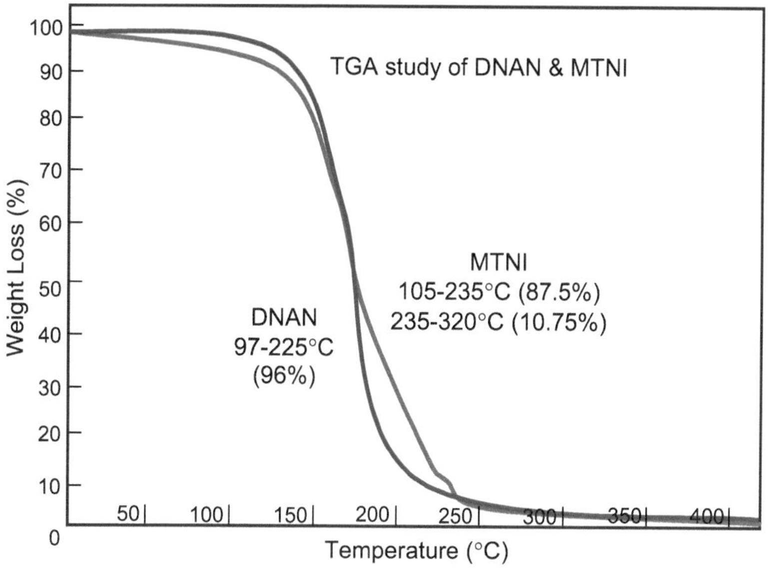

Figure 10.6
TGA of DNAN and MTNI.

possible decomposition mechanism of HEMs when combined with other techniques such as GC/mass spectrometry.

10.4.4 Simultaneous Thermal Analysis

STA is a coupled technique in which a sample is heated in a programmed fashion and the thermal events are recorded simultaneously by using two different techniques (e.g., DSC and TGA or DTA and TGA). Simultaneous recording of TGA and DTA/DSC can provide better insight into the probable physicochemical mechanism involved during the decomposition/oxidation of the sample.

10.5 Sensitivity Tests of HEMS

The sensitivity of an HEM is its response to an external stimulus such as impact, friction, shock, and electrostatic discharge. Because the utmost care must be taken at every stage in the field of HEMs (e.g., formulation, processing, transportation, and storage), and one or more stimuli, as mentioned above, may be encountered by HEMs during any of these stages, the concerned HEM must be thoroughly tested for its sensitivity to most or all of the above stimuli. Because the sensitivity of HEMs varies depending on the nature of the HEM and the stimulus involved, in each case a reference HEM is taken to assess the relative sensitivity of a given HEM. For example, composition exploding (CE), which is 2,4,6-trinitrophenyl methylnitramine (Tetryl), is taken as a reference explosive with an impact sensitivity of 70, and, on this scale, the relative impact sensitivity of other explosives is assessed. A brief mention is made in the following about the various types of sensitivity tests commonly used in the field of HEMs.

10.5.1 Impact Sensitivity

Impact sensitivity is the ability of a substance to withstand a sudden blow without decomposing or igniting when impacted by the fall of a 2-kg hammer from a specific height. A 2-kg weight is released to impact the sample kept on an anvil and the result is noted. The impact distances are determined, and the results are analyzed by the Bruceton staircase method. In this method, one measures the 50% initiation level (the height at which 50% of the samples will detonate), compares it with that of a standard explosive (CE taken as 70), and reports it as the figure of insensitiveness (F of I). This height is called the median drop height of the sample.

$$\text{Figure of Insensitivity of the sample (F of I)} = \frac{\textit{Median dropheight of sample}}{\textit{Median dropheight of reference}} \times \text{(F of I of standard)}$$

For example, if a high explosive and CE (standard) give median drop heights of 60 and 80 cm, respectively, in an experiment, the F of I of the explosive is given as

$$\text{F of I} = \frac{60}{80} \times 70 = 52.5$$

10.5.2 Friction Sensitivity

Friction sensitivity is the measure of sensitivity to initiation of an HEM to relative frictional motion between two objects in contact. The sensitivity of an explosive is determined by subjecting a thin layer of explosive on a predetermined spot to a certain load through a loading arm and allowing the explosive to undergo standard frictional movement. The range of load for primary explosives is 10–1000 g and for other explosives is 0.5–36 kg. The observation of any effects (e.g., smoking, burning, or exploding) at a minimum particular load is taken as the sensitivity to friction of an explosive to that load.

10.5.3 Spark Sensitivity

Repeated contact and separation of two dissimilar materials (one of them is insulator material) lead to accumulation of electric charge, which often gets discharged through lower potential materials in the surrounding. Hence, it is vital to measure the electrostatic sensitivity of explosive stores to understand the hazards associated with that particular material. These data help in designing the safety measures required for reducing the electrostatic hazards in various operations such as mixing, sieving, handling, storage, etc. This assessment can be done by exposing a known quantity of a particular sample from low discharge energy to high discharge energy in an incremental fashion until the sample gets ignited.

Suggested Reading

[1] J. Yinon, S. Zitrin, Modern Methods and Applications in Analysis of Explosives, John Wiley and Sons, 1996.
[2] U. Teipel, Energetic Materials Incorporation of Particular Components with Specialised Properties Allows One to Tailor the End Product's Properties, Wiley-VCH Verlag GmbH, 2004.
[3] W. Kemp, Organic Spectroscopy, second ed., Macmillan, 1987.
[4] J.H. Michael, Modern Spectroscopy, fourth ed., Wiley, 2004.
[5] J.P. Agrawal, R.D. Hodgson, Organic Chemistry of Explosives, first ed., Wiley, 2007.
[6] J. Akhavan, The Chemistry of Explosives, third ed., Royal Society of Chemistry, 2011.

Questions

1. What is the difference between the characterization and evaluation of an HEM? Give an example.
2. What is the basic principle of chromatography?
3. Why is HPLC a preferred technique in the separation and analysis of HEMs?
4. Why are many HEM molecules active in the UV region?
5. Why do we get three clear, different NMR spectral peaks in ethyl alcohol?
6. Why are thermal analysis techniques very important in the field of HEMs?
7. What is the major difference between DTA and DSC?
8. What are the uses of the TGA technique in the field of HEMs?
9. What are the different types of sensitivity of HEMs and how are they measured?

CHAPTER 11

HEMs: Trends and Challenges

11.1 Introduction

From the first experiments with gunpowder and fireworks to the latest ultra-powerful high-explosive, octanitrocubane (ONC), or caged nitramine class of explosives such as CL-20, man has sought to unleash the force of chemical explosives in more powerful and controlled ways. However, the rapidly changing technology presents some tremendous opportunities and pitfalls. Now more than ever, success on the battlefield is dependent on the rapid access to information and the ability to act on that in a timely manner. The current trends in the field of high energy materials and the future challenges are discussed briefly in this chapter.

11.2 Primary Explosives

11.2.1 Problems

Mercury fulminate and lead azide are the foremost primary explosives, and they gained prominence in military ammunitions and civil applications. They ruled the world of initiatory compounds for more than 70 years. Despite being good in performance, they suffer from certain inherent drawbacks, such as hydrolytic instability, incompatibility with copper or its alloys (commonly used for encapsulation of primary explosive formulations), and high friction sensitivity. Globally, research and development is triggered for the development of potential primary explosives with figures of insensitivity greater than 20, which are then less prone to accidental initiation during storage, transport, or handling of the finished ammunitions and are stable and compatible with copper and its alloys.

11.2.2 Solutions

Lead-free coordination compounds are the choice of tomorrow in view of their additional advantage of being ecofriendly. Another desirable attribute of this class of compounds is the presence of almost stoichiometric fuel and oxidizer moieties. These compounds have been known for some time, but their applications in primary explosives were recently realized. Two important energetic coordination compounds are nickel hydrazine nitrate (NHN) and bis-(5-nitro-2H-tetrazolato-N^2) tetramine cobalt(III) perchlorate (BNCP).

Demystifying Explosives: Concepts in High Energy Materials. http://dx.doi.org/10.1016/B978-0-12-801576-6.00011-2

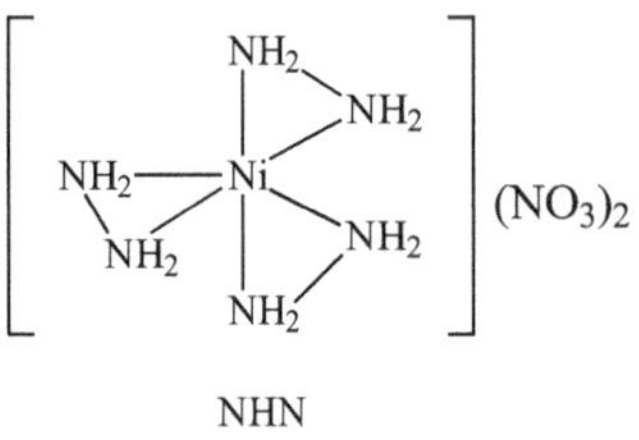

NHN

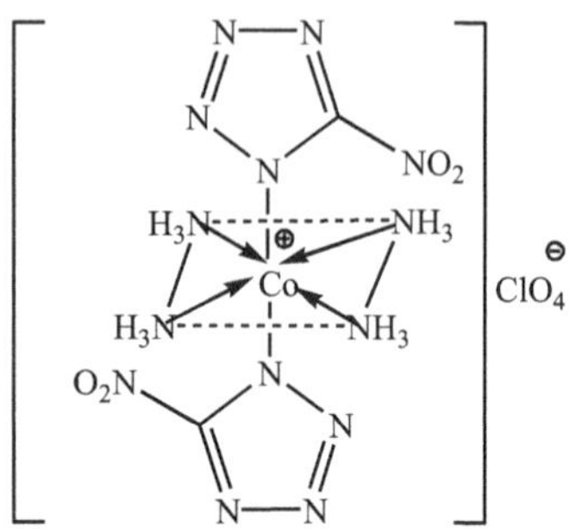

BNCP

Apart from these, research is focused on the development of energetic coordination compounds without deleterious metal ions such as Co, Ni, and anions such as ClO_4^- and NO_3^-.

11.3 High Explosives

High explosives are the major components of any weapon system. Today, the candidate molecules available for warhead applications are very limited. Any researcher always looks for better materials than the existing benchmark candidates. In this regard, any high-explosive scientists and technologists always look for better materials than HMX.

Three categories of importance under high explosives need to be mentioned here:

1. High-density, high-VOD (velocity of detonation) explosives, which, because of their superior power (due to high VOD and hence high detonation pressure) and volumetric efficiency (due to high density), hold great promise to impart enormous lethality to ammunitions of the future.
2. Insensitive explosives (or low-vulnerability explosives) assume their importance in view of the safety involved during transportation and storage and to prevent accidental initiation (e.g., due to enemy fire) during a battle. There are often battlefield situations in which the power of the explosives can be slightly compromised to ensure low vulnerability of the ammunition and avoid any disaster.
3. Thermally stable explosives (sometimes referred as heat-resistant explosives) play a vital role when a high-temperature environment may adversely affect the performance (or worse, prematurely initiate) of the explosive. Examples of such an environment are the explosives filled with warheads of supersonic missiles in which aerodynamic heating is involved and explosives are used in oil well exploration. Similar to the case of insensitive explosives, thermal stability is imparted slightly at the cost of the power of the explosives. As one can expect, imparting all three qualities of high power, high insensitivity, and high thermal stability in an explosive is impossible. To give an

example, a certain amount of thermal stability is imparted in an aromatic polynitro compound by converting one or more $-NO_2$ groups into $-NH_2$ groups (so that inter/intramolecular hydrogen bonding between $-NO_2$ and $-NH_2$ groups form a sort of a matrix in the explosive, introducing higher thermal stability). However, sacrificing one or more energetic $-NO_2$ groups to introduce nonenergetic $-NH_2$ groups results in the decreased power of the explosive. The choice of 1, 2, or 3 depends on the type of ammunition in which they are to be used.

11.3.1 High-Density, High-VOD Explosives

The challenges faced by an explosive chemist to achieve their goal of synthesizing an explosive more powerful (higher VOD) and with higher density are amazingly high. They must visualize a nonaromatic, polycyclic "caged" compound containing energetic groups such as $-NO_2$ or $N-NO_2$. Visualization is just the first step; they must assess its energetics by various empirical and quantum mechanical methods to calculate its expected VOD and density, if at all it is synthesized. Even if there are a 100 candidate molecules to start with, the chemist may ultimately end up with a few or even none. This is because their choices are ruthlessly narrowed down when most (if not all) of the candidates are rejected in terms of huge costs and hazards because of extreme sensitivity or instability of the final explosive or even intermediates.

If a new explosive with marginally better properties than HMX is going to be 100 times costlier than HMX, then no one is going to use it as ammunition. Or, if the same explosive is very sensitive to mechanical shock or has a poor thermal stability, then it holds no promise. This is the main reason why new powerful and promising explosives are made after long intervals of several decades. RDX was first made in the year 1899. Its higher homolog, HMX, was later made only in the year 1943 after a gap of 44 years.

Under this category, the only two high explosives exceeding the performance of HMX are hexanitrohexaazaisowurtzitane also called China Lake-20 (CL-20) and ONC. The synthesis and subsequent manufacture of CL-20 were established only in a few selected countries. The cost of CL-20 is much higher than that of HMX because it involves quite a few steps using very costly catalysts. Moreover, its sensitivity to impact and friction is more than that of RDX and HMX (see Table 11.1). Despite the cost and sensitivity factors, CL-20 holds great promise in futuristic warheads and its role as an ingredient in low-signature propellants because of its better oxygen balance than RDX and HMX.

Another interesting candidate in the series is ONC (see Table 11.1). ONC was synthesized in 1999 using a multistep synthesis approach. No reports appeared in the literature after its invention because of the difficulties in the preparation methods. Research and development efforts have to be focused for a viable and simple method for the preparation of ONC.

Table 11.1: Comparison of RDX, HMX, CL-20, and ONC.

Parameters	RDX	HMX	CL-20	ONC
Structure	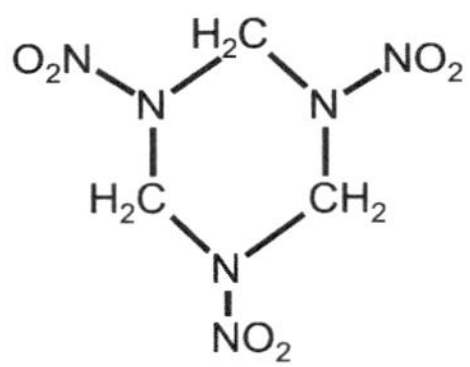	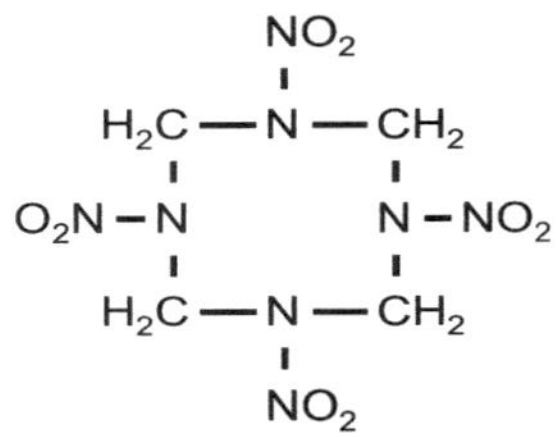	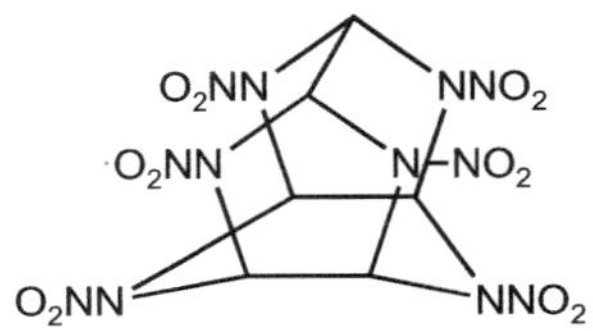	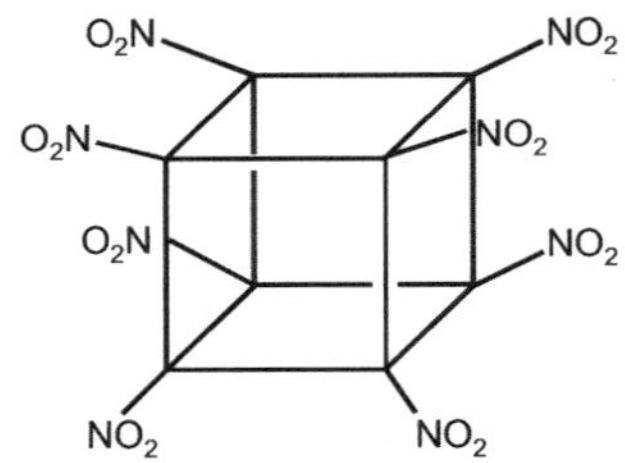
Year first made	1899	1943	1987	1999
Density (g/cm^3)	1.81	1.91	2.04	2.1
Oxygen balance (%)	−21.6	−21.6	−11	0
VOD (m/s)	8800	9100	9400	9800
Sensitivity to impact (h50%, cm)	46	38	24	Not reported
Friction (kg)	16	14	8	Not reported

In modern ordnance there is a strong requirement for explosives having thermal stability and mechanical insensitivity coupled with better explosive performance. Under normal conditions of use, munitions filled with conventional explosives are safe and effective to provide desired military capability. However, they violently respond to unintentional initiation such as getting exposed to enemy fire. In general, high explosives, gun propellants, and rocket propellants are sensitive to heat, mechanical shock, fire, and mechanical impact by bullets or fragments. Such secondary effects may lead to the violent initiation of the ammunitions and result in huge loss of men and materials on a battlefield.

11.3.2 Insensitive High Explosives

To avoid this kind of collateral damage to weapons, a new concept called "insensitive munitions" (IMs) has emerged. IMs are munitions designed to minimize the consequences of an accidental initiation without compromising the expected performance. IMs are designed such that they are difficult to be ignited accidentally or, in the case of an accidental initiation, it does not result in any detonation or mass fire.

The synthesis of nitrotriazoles as insensitive energetic materials has received a great deal of attention in the past 20 years. The most studied nitrotriazole explosive is 3-nitro-1,2,4-triazole-5-one (NTO). NTO is currently being widely investigated in main charge warhead filling for IMs. It is used in cast-cured, pressed, and sheet explosive formulations.

Another new insensitive energetic material with promising properties is 1,1-diamino-2,2-dinitroethylene (FOX-7). In recent years, much interest has been devoted to RDX in another form (i.e., reduced sensitivity RDX (RSRDX) or insensitive RDX), which, when incorporated in cast-cured plastic-bonded explosive formulation, can confer reduced shock sensitivity as measured through the gap test. RSRDX is reported to possess improved crystal density and fewer crystal defects with smooth surface morphology, and these factors seem to reduce the mechanical sensitivity of RDX. These features also impart reduced vulnerability toward shock initiation. The preparation method of RSRDX is not disclosed in open literature because of its strategic importance. The properties of some of the insensitive explosives are given in Table 11.2 in comparison with some thermally stable explosives.

11.3.2.1 Is TNT Suitable for IM?

In the area of melt-cast explosives, trinitrotoluene (TNT) is the well-known explosive used in all possible ammunitions since World War I. The advantage of TNT is that it melts at 81 °C and can be cast alone or in combination with other ingredients such as RDX, aluminum, and ammonium perchlorate (AP) into various desired shapes. Although the performance is lower, TNT was well accepted by ordnance communities because of the above advantages. However, TNT and TNT-based ammunitions pose long-term health

Table 11.2: Performance comparison of insensitive and thermally stable explosives.

Name of HEM	Structure	Oxygen Balance (%)	Density (g/cm³)	VOD (m/s)	Impact Insensitivity ($h_{50\%}$, cm)	Friction Insensitivity (kg)
		Insensitive High Explosives				
NTO	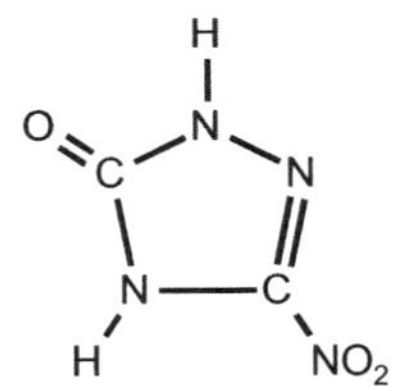	−24.6	1.93	8564	93	>36
FOX-7	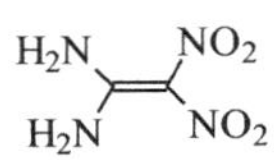	−21.61	1.88	9090	126	>36
		Thermally Stable Explosives				
TATB	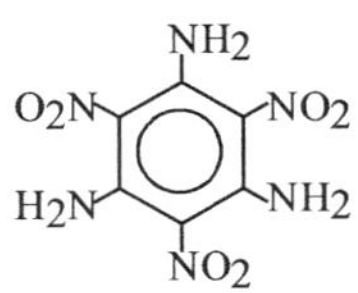 TATB Decomposition temperature: 376 °C (T_{max})	−55.78	1.94	8108	>177	>36

LLM-105	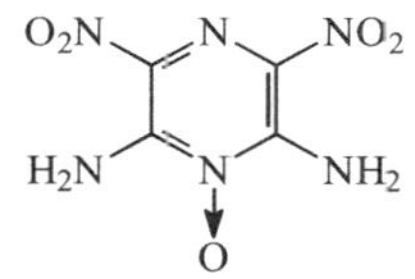 LLM-105 Decomposition temperature: 342 °C	−37	1.91	8560	117	>36
TACOT	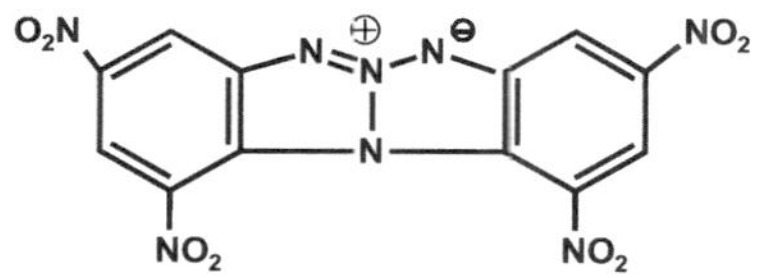 Decomposition temperature: 403 °C (T_{max})	−74	1.82	7060	68	>36

hazards to workers. Moreover, formulations containing TNT exude during storage. In addition, numerous problems involving TNT are noted during the melt-pour process, including high volume change from liquid to solid, supercooling, irreversible growth of crystals, and unpredictable sensitivity.

The main problem with TNT as a filling ingredient for modern projectiles is that it behaves violently if subjected to an accidental stimulus, such as being involved in a fire attack by enemy gun fire. The efforts pursued to make TNT safer ended in failure. Likewise, TNT-based ammunitions have failed all IM tests.

A promising compound to replace TNT as a melt-cast explosive is 2,4-dinitroanisole (DNAN) with a melting point of 94 °C. DNAN-based compositions with other ingredients such as RDX, aluminum, and AP successfully passed IM tests, which are internationally accepted. However, DNAN is inferior in performance to TNT. Hence, research and development efforts realized another potential compound, namely *N*-methyl-2,4,5-trinitro imidazole (MTNI), which melts at 82 °C. This compound possesses good thermal stability, impact insensitivity (50–70 cm), and better explosive performance than DNAN and TNT. However, the main problem with MTNI is its low yield in preparation.

Likewise, another high-performance melt-cast explosive realized in the last decade was 1,3,3-trinitroazetidine (TNAZ), which is a strained nitramine compound with a higher melting point (102 °C). This compound suffers in many aspects, such as a multistep cumbersome synthesis approach, high volatilization, exorbitant cost, and health hazards to the workers in the vicinity. Hence, not much seriousness was shown in the development of TNAZ-based ammunitions. Important melt-cast candidates are presented in Table 11.3.

11.3.2.2 Thermally Stable Explosives

Warhead fillings of modern weapons are expected to function under various environmental conditions. Improved thermal stability of explosives in such warheads increases the shelf life of munitions. Their heat resistance decreases their vulnerability to accidental initiation. The development of explosives for space programs, applications in oil well exploration, transportation of munitions by supersonic aircrafts, etc., have resulted in the need for thermally stable explosives. The properties of some of the thermally stable explosives and which hold promise are given in Table 11.2, along with a few insensitive explosives.

11.4 Propellants

11.4.1 Ecofriendly Oxidizers

In today's war scenario, a rocket propellant has the dual requirement of high performance and low signature (smoke). The major disadvantages of rocket propellant formulations

Table 11.3: Potential melt-cast explosives with performance.

Parameters	TNT	DNAN	MTNI	TNAZ
Structure	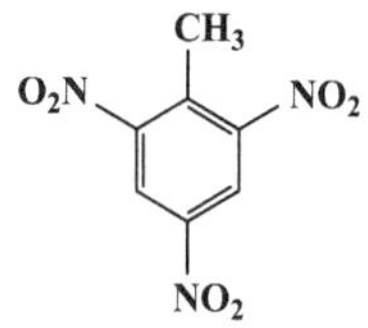	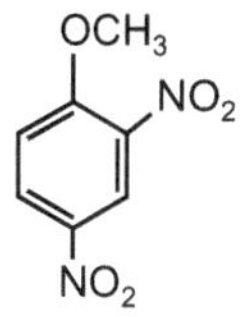	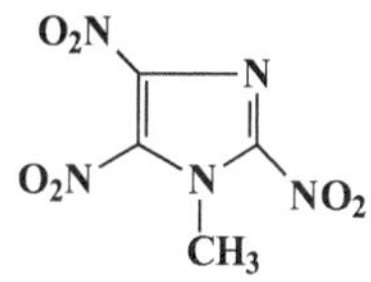	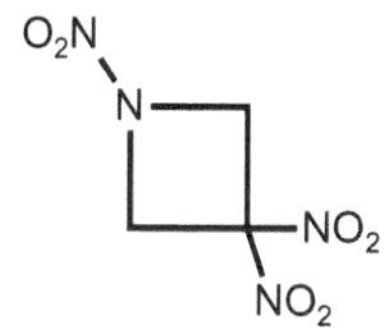
Melting point (°C)	80.8	94	82	102
Density (g/cm^3)	1.65	1.55	1.76	1.84
Oxygen balance (%)	−74	−97	−25	−16.6
VOD (m/s)	6900	6800	8000	9000
Sensitivity				
Impact insensitivity (h50%, cm)	>170	>170	62	45–47
Friction insensitivity (kg)	>36	>36	>36	>36

with AP are that they produce huge signature because of the emission of hydrogen chloride (HCl) gas and they pollute the environment. Another drawback of AP is that it inhibits the functions of the thyroids of personnel involved in large-scale AP processing over a period of time. Therefore, a large amount of money is still being spent on the development of ecofriendly and low-signature oxidizers that will replace AP.

One such oxidizer is ammonium dinitramide (ADN). It is an inorganic oxidizer and was first made by a Russian scientist in the late 1970s, and the preparation details were kept under a high order of secrecy because of its strategic importance. ADN-based propellants offer high specific impulse with no secondary smoke because of the absence of HCl. The major application of ADN is that it can replace today's workhorse oxidizer AP in rocket propellants. The synthesis of ADN has been reported widely in the literature using various synthesis approaches. The main problem of ADN is that it is very sensitive to moisture. ADN readily absorbs moisture and rapidly decomposes. Stabilization of ADN is a critical issue and has been done through prilling or by coating techniques.

Hydrazinium nitroformate (HNF) is another ecofriendly energetic oxidizer for solid rocket propellants. The drawback of HNF is that it is very sensitive to mechanical stimuli, particularly to friction because of the sharp needle-shaped crystals. Hence, it is not possible to directly use it in propellant formulations. To overcome this problem, desensitization of HNF is necessary to process HNF-based propellants. Desensitization of HNF calls for elaborate trials involving the addition of proper ingredients at the time

Table 11.4: Performance comparison of various oxidizers.

Parameters	AP	ADN	HNF
Structure	NH_4ClO_4	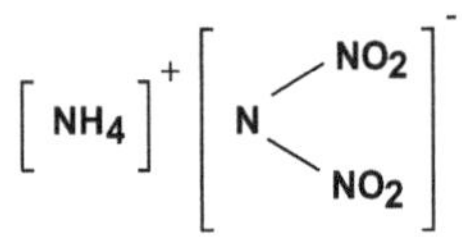	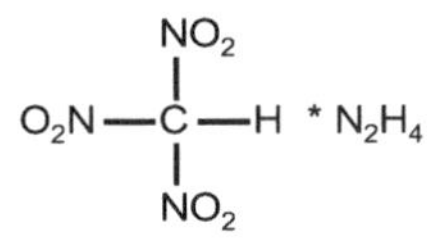
Melting point (°C)	452	92–93	115
Density (g/cm^3)	1.9	1.8	1.9
Oxygen balance (%)	34	26	13

of crystallization to modify its morphological characteristics. HNF can find applications in futuristic low-signature, high-performance green propellants in place of AP once the sensitivity issue is resolved. All said and done, despite the drawbacks of AP, there is a long way to go to replace AP because of its excellent oxygen balance, ease of preparation in large scale, and low cost. Table 11.4 gives a comparison of AP, ADN, and HNF.

11.4.2 Metallic Fuels

Composite rocket propellants use metallic powders as fuels. Most of the modern composite solid propellants contain finely powdered metallic fuels such as aluminum. They increase the chemical energy of the propellants by increasing the combustion temperature due to large thermochemical energy output.

For several decades, aluminum has been the choice in propellant formulation because of its reasonably good thermochemical energy output, easy availability, nontoxicity of combustion products (mostly aluminum oxide), and low cost. The search is on to replace aluminum by metals that are more energetic and dense to boost the performance of the propellants. However, the alternative metallic fuels pose problems such as toxicity of products, combustion instability, high cost, etc. For example, boron is an alternative metallic fuel, but it is difficult to ignite/burn. Beryllium is energetically more favorable than aluminum, but it produces highly toxic products on combustion; therefore, it is not acceptable.

Zirconium has attractive properties in terms of density and energetics, but it is very hazardous in view of its pyrophoric nature (easy ignitability in the presence of air). In addition, there are certain metal hydrides that are also being tried as fuels in advanced propellant formulations in view of their attractive energetics. Lithium aluminum hydride is toxic and dangerous to handle because it may ignite and violently burn. It is incompatible with water, alcohols, ammonium hydroxide, etc.; however, it acts as a high-energy fuel. Likewise, toxicity and sensitivity to initiation by mechanical shock ruled out the use of magnesium hydride and lithium borohydride. Similar to AP in the case of oxidizers, it will take quite some time to completely replace aluminum as a fuel in large-scale propellant processing.

11.4.3 Energetic Binders

Binders are typically cross-linkable polymers (or sometimes called prepolymers) added in propellant formulations to bind the solids (oxidizer, fuel, additives) together with a plasticizer and to enhance the mechanical properties of the composition. For several decades, the choice of binders (which also act as nonmetallic fuels) for rocket propellants has been based on hydrocarbons such as polybutadiene. Carboxyl-terminated polybutadiene (CTPB) and hydroxyl-terminated polybutadiene (HTPB) are popular among them. Although the repeating unit of the polybutadiene chain

$$\left(CH_2-CH=CH-CH_2 \right)$$

releases a good amount of heat on combustion, scientists have been working on the introduction of energetic functional groups, such as $-NO_2$, $-NO_3$, and $-N_3$, in the backbone of the polymeric binder (or sometimes as pendent groups attached to the backbone) to enhance the energy output during the propellant combustion. However, this does affect the easy processability of the propellant because the viscosity of the binder substantially increases due to the introduction of such energetic groups in the polymer backbone.

Some of the candidate polymers containing energetic groups such as $-N_3$, $-NO_3$, etc., are based on a polyethylene oxide

$$HO\left(CHR-CH_2-O \right)_n H$$

backbone (e.g., glycidyl azide polymer (GAP) and polyglycidyl nitrate (PGN) or a polypropylene oxide

$$HO\left(CH_2-CR_2-CH_2-O \right)_n H$$

backbone (e.g., poly-3,3-bis(azidomethyl) oxetane (polyBAMO) and poly-3-nitratomethyl-3-methyloxetane (polyNIMMO). Their molecular structures are shown in Table 11.5.

At times, some of these polymers (the viscosities of which are quite high) are copolymerized with nonenergetic (low-viscosity) polymers such as polytetrahydrofuran

Table 11.5: Physicochemical properties of some energetic binders.

Polymer	Structure	Density (g/cm³)	Oxygen Balance (%)	Glass Transition Temperature (°C)
HTPB	$HO{-}(CH_2{-}CH{=}CH{-}CH_2)_n{-}OH$	0.92	−324	−65
GAP	$HO{-}(CH(CH_2N_3){-}CH_2{-}O)_n{-}H$	1.3	−121	−50
PGN	$HO{-}(CH(CH_2ONO_2){-}CH_2{-}O)_n{-}H$	1.39	−61	−35
PolyBAMO	$HO{-}(CH_2{-}C(CH_2N_3)_2{-}CH_2{-}O)_n{-}H$	1.3	−124	−39
PolyNIMMO	$HO{-}(CH_2{-}C(CH_2ONO_2)(CH_3){-}CH_2{-}O)_n{-}H$	1.26	−114	−25

viz., $HO{-}(CH_2{-}CH_2{-}CH_2{-}CH_2{-}O{-})_n{-}H$ for improving the processability. Table 11.5 compares the properties of some energetic binders.

11.4.4 Thermoplastic Elastomers

All of the polymers discussed so far are chemically cross-linked by a curing agent; hence, they have a certain amount of rigidity. They come under the category of thermosetting polymers and cannot be reprocessed. Thermoplastic elastomers (TPEs) are popular choices when one wants to process propellant compositions that can be reprocessed and that are easily disposed. TPEs contain macromolecules, each having a backbone containing "hard" (glassy) segments (e.g., aromatic rings) and soft (rubbery) segments (e.g., a polybutadine moiety). Only physical cross-links give the polymer a physical or structural integrity, and they start disappearing near the melting point (like untying a complex knot). During cooling, these cross-links reappear. The thermoplastic and elastomeric nature of TPEs has been exploited in using them for processing the propellant by extrusion methods. Some of the TPEs are also being tried for extrudable gun propellants for the same reason. One of the great advantages of TPE-based ammunitions is the ease of demilitarization (i.e., the ammunitions can be easily disposed by the process of melting).

11.4.5 Energetic Plasticizers

Plasticizers are low molecular weight liquids added to a polymer at the time of processing. The plasticizer molecules penetrate through the interstices between the long chains of the polymer and get linked to the polymer chain through weak physical bonds, thereby decreasing the interchain attractive forces in the polymer. This gives a "greasing" effect so that the polymer chains can slide among themselves. Thus, the plasticizer gives flexibility to the finished polymer. In addition, during the polymer processing, the plasticizer reduces the viscosity of the mix, thereby improving the processability. Many popular plasticizers used in the propellant industry have been nonenergetic, such as phthalate esters and a few aliphatic ones. The conventional energetic plasticizer well known in double-base propellants is nitroglycerine (NG). NG is an excellent plasticizer of nitrocellulose and it contains energetic $-ONO_2$ groups. However, NG is highly sensitive to impact; hence, its use as a plasticizer is limited.

Modern research replaces the nonenergetic plasticizers with an array of energetic nitrate esters such as butanetrioltrinitrate (BTTN), triethylene glycoldinitrate (TEGDN), butanenitratoethylnitramine (BuNENA), bis-(2,2-dinitropropyl) acetal/formal (BDNPF/A), low molecular weight GAP (GAP plasticizer), and trimethylolethane trinitrate (TMETN). These plasticizers may be used independently or in combination with other plasticizers. Table 11.6 compares the properties of some energetic plasticizers.

Apart from oxidizers, fuels, and binders, intense research and development has been going on for choosing better materials for other propellant ingredients such as burn rate modifiers and other process aids. As an example, to improve the solid loading characteristics, efforts are on to replace the conventional burn rate catalysts that are solids (e.g., iron(III) oxide or copper chromite) with liquid ones (e.g., ferrocene-based oligomers).

11.5 Polynitrogen Cages: Promising a Revolution in Future HEMs?

HEM scientists have ambitious plans for the future. On the basis of simple logic and extensive quantum mechanical calculations, the ultimate target molecules that will be the HEMs of the future must be those that have very high positive heats of formation, high densities, and very large heat release, resulting in very high VODs and detonation pressures if used as high explosives and very high I_{sp} values if used as propellants. If we

Table 11.6: Physicochemical properties of some energetic plasticizers.

Plasticizers	Structure	Density (g/cm^3)	Oxygen Balance (%)
NG	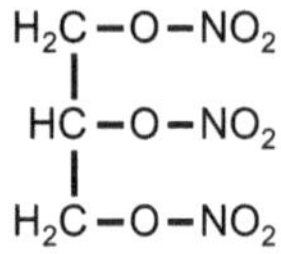	1.59	+3.5
BTTN	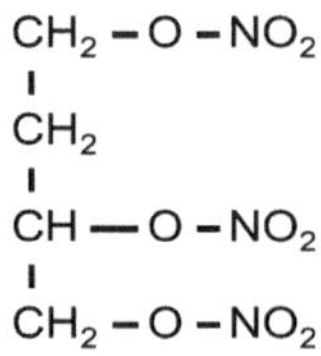	1.52	−16.6
n-BuNENA	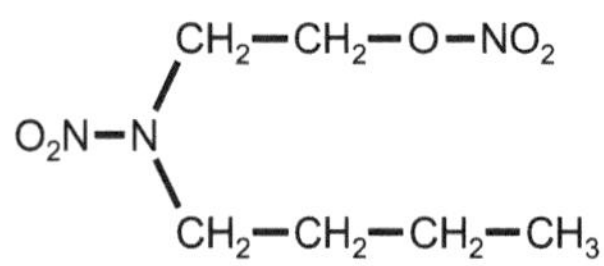	1.20	−104
TMETN	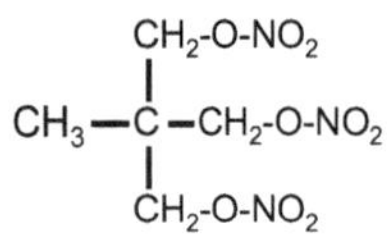	1.48	−34
BDNPF/A	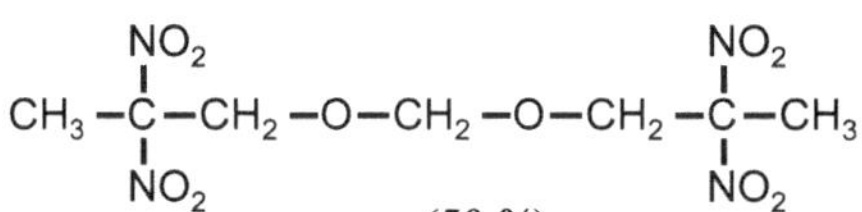 (50 %) 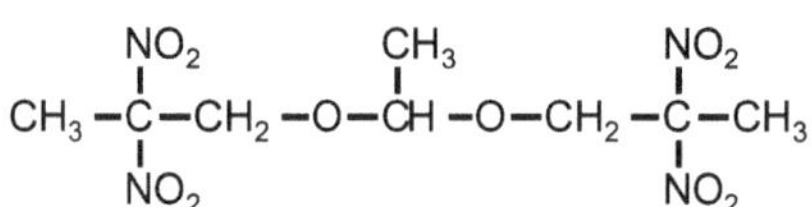 (50 %)	1.39	−51

compare RDX, HMX, CL-20, and ONC (in the same order), then we find that their densities, heats of formation, and VOD values significantly increase. One can visualize that as we go from RDX to ONC, the ring strain in the molecule increases. This strain and the nature of nitrogen bonding remarkably contribute to the positive values of heats of formation and the energetics of the molecules. Extending this picture further, the hopes are pinned on those molecules that contain only nitrogen atoms in a strained ring structure. For example, imagine an N_8 molecule in which eight nitrogen atoms occupy the eight corners of a cube. The bond angle in this molecule becomes 90°, which is far less than 109°, a comfortable bond angle for the nitrogen compounds in which the nitrogen atom is bonded to three other atoms. Therefore, one can expect a very great degree of strain experienced by the cubic structure of N_8, resulting in very high values of the heat of formation for the molecule. Such a molecule will be a dream molecule for any HEM scientist because when an N_8 molecule decomposes to give four molecules of nitrogen, the energy released will be stunningly high.

However, the problem is the huge challenges involved in their synthesis. Some years back, when polynitrogen compounds such as $Mg(N_5)_2$, $N_5^+SbF_6^-$, and $N_5^+SnF_6^-$ were made, it spurred the ambition of HEM scientists for planning the synthesis of polynitrogen molecules such as N_8 and N_{60}. N_8 and N_{60} will theoretically have the VOD values of 14.9 and 17.31 km/s, respectively (for HMX, it is 9.1 km/s), and heat of formation values of 407 and 546 kcal mol^{-1}, respectively (for HMX: 28 kcal/mol). However, the challenges involved in the chemistry of their synthesis are intimidating. Although some limited reports are available on the synthesis of some polynitrogen compounds such as those based on N_5^+ in the literature, it is going to be a very long and arduous journey for the HEM scientist to reach these goals.

Suggested Reading

[1] J.P. Agrawal, R.D. Hodgson, Organic Chemistry of Explosives, first ed., Wiley, 2007.
[2] J. Ledgard, The Preparatory Manual of Explosives, third ed., 2007.
[3] T.M. Klapötke, High Energy Density Materials Series: Structure and Bonding, first ed., Springer, 2007.
[4] R. Meyer, J. Kohler, Explosives, VCH Publishers, Germany, 1993 (Encyclopaedia – handy for referencing).
[5] D.H. Liebenberg, et al. (Eds.), Structure and Properties of Energetic Materials, Materials Research Society, Pennsylvania, USA, 1993.
[6] J. Akhavan, The Chemistry of Explosives, third ed., Royal Society of Chemistry, 2011.
[7] N. Kubota, Propellants and Explosives Thermochemical Aspects of Combustion, 2007.

Questions

1. What is the necessity of lead-free initiatories?
2. Initiatory compounds should be sensitive. Justify the statement.
3. What is meant by coordination compounds? Name any two coordination compounds used for primary explosive purposes.
4. What is hydrogen bonding? How is it useful in achieving insensitivity/thermal stability of explosives?

5. What are the advantages of melt-cast explosives?
6. Explain the meaning of demilitarization?
7. Why is the viscosity of a polymer increased while introducing pendent groups?
8. What are the potential polymers that might replace HTPB in the future?

CHAPTER 12

HEMs: Constructive Applications

12.1 HEMs Have Shaped Our World

Ever since Alfred Nobel invented dynamite about 140 years ago, the world has undergone an incredible transformation. Population has increased tremendously and so have the global materialistic demands. Technology has, in different fields, grown to amazing levels. The march toward better technologies and better products goes on unrelentingly. There have always been remarkable milestones in the history of development of science and technology, and the achievement of each milestone changed the very face of our life on this earth. Some of the milestones achieved in the field of HEMs have literally shaped our world. It is an undeniable fact that explosives have wreaked untold havoc and horrors (and are still wreaking sporadically!) in the guise of a number of wars since last two centuries or more. But it is also an undeniable fact that the explosives or generally HEMs have shaped the world what it is today. The object of this chapter is to highlight this "other side of the coin," namely, the role of HEMs for constructive purposes.

Advancement in science and technology would have been almost impossible but for the fact that HEMs paved the way to easily tap the earth's resources. Among many constructive applications of HEMs, the following stand out undoubtedly.

12.1.1 Mining and Quarrying

Coal mining has been feeding the vast energy requirements of the mankind, although today we are nervously aware that this fossil fuel will not last forever. All metals and minerals (which play important roles in materials and equipments of everyday use, be it toothpaste or talcum powder, medicine, cosmetics or color TV, and computer chips) have made our life richer and more comfortable.

12.1.2 Construction

Amazing augmentation of infrastructure throughout the world has changed the very face of the earth in the last many decades and is still continuing with unabated speed. Construction of huge multistoried structures, roads, tunnels, bridges, etc. has been contributing greatly to the economy of many nations.

Demystifying Explosives: Concepts in High Energy Materials. http://dx.doi.org/10.1016/B978-0-12-801576-6.00012-4

12.1.3 Oil Well Perforation

Today, there are frenzied efforts to find alternative sources of energy driven by the fear of exhausting all the fossil fuel resources. Still, the fact remains that the oil, aptly called black gold, is the lifeline of our existence today. Imagine today's world without oil just for a week: everything would come to a grinding halt!

It is not hard to guess that the very basic requirement for the above vital activities is HEMs. In the past, humans had also been mining out coal, iron, copper, and other minerals. However, after the invention of dynamite and subsequent civil explosives, there was a 100-fold increase in their production. So was the case with quarrying. There was a tremendous increase in the production of cement and concrete and huge leap in the construction activities. Between American Civil War (1776) and end of the World War II (1945), no single engineering tool surpassed the achievement of dynamite. Today, Explosives Engineering is a specialized field and is undergoing continuous improvement (some of the basic aspects of Civil Explosive have been covered in Chapter 4). It is a multidisciplinary field that involves chemistry of explosives, detonics, structural engineering, etc.

In the following section, let us briefly see the application of HEMs in certain other not-so-common areas.

12.2 Controlled Demolition

Imagine a situation like this: A thirteen-story building that has outlived its utility needs to be demolished. The hitch is that there is a massive hospital complex with even an organ transplant facility in close proximity apart from other high-rise structures. Conventional methods of demolition using hammering, bursting, etc. will not only take enormous time, labor, money, etc., but will also involve a host of problems like traffic dislocation in the nearby area, continuous emission of noise, and enormous amounts of dust and debris. Such a process is very likely to cause serious pollution problems and potential infection to the patients in the adjoining hospital complex. Moreover, the conventional methods of demolition call for a large number of machineries like cranes, which pose severe problems of space and logistics while demolishing a structure in a congested area. Actually, the above situation was faced by an Irish hospital complex a few years ago, and that is when the controlled demolition by explosives became quite handy.

12.2.1 Explosion or Implosion?

We know that if we want to blast a multistoried structure by explosives in the conventional way, the shockwave created as well as the flying debris of steel and concrete will wreak unimaginable havoc on life and property all around. But in the controlled demolition by explosives, it is necessary to *implode* the building so that it collapses down into its footprint. An implosion can be defined as an event where something *collapses inward*, because of the external/atmospheric pressure. For example, if you pump out the air out of a thin glass vessel, it might implode. Strictly speaking, controlled demolition of a building is not truly an implosion: atmospheric pressure does not pull or push the structure inward. Here, the explosives are used to weaken the supporting structures like columns/pillars, thus allowing the gravity to pull the structure down by the virtue of its own weight. The resultant huge piles of debris are not "laid out," but they fall very close to the foundation of the structure.

It you have a four-legged table and you remove two legs from one side, the table will fall over. You can control the direction of fall by choosing the appropriate two legs that are to be removed. A large building generally has many "legs" or columns that support it. In an implosion, first you remove the columns from within the building, thus causing the initial collapse to start from that point. The initial collapse of the inner columns helps to drag the structure down toward the center.

Remember the catastrophic collapse of the World Trade Center structures in the infamous 9/11 attacks at New York? Two of the tallest buildings in the world collapsed just vertically without causing much damage to the nearby structures. It is probable that the high temperature flames made the supporting structures give way. The rest of the job was done by gravity.

12.2.2 Step-by-Step

The actual process of implosion may take place in less than 60 s. However, prior to the implosion, on-site preparation operations will take several weeks to complete. Key structural supports are identified and exposed by removing interior, non-load bearing walls and piping. Small diameter holes will be drilled at specific locations to act as explosive receptacles equipped with internal non-electric timing devices that will fire on queue. Some of the important measures to be taken include the minimization (total elimination is not possible) of dust production and vibration.

The extent to which the nearby buildings or facilities will be affected by dust depends on the wind speed and direction at the time of implosion. Dust-producing materials from the building such as dry wall plaster, ceramic tiles, etc. are to be removed initially. The implosion will be designed to minimize the amount of vibration. Other precautions include closing of windows/doors/exhaust fans/air conditioners, etc. in the neighborhood during the implosion followed by some period.

Explosive (or implosive) demolition of buildings is safe, cheap, and quick, but caution! This needs to be carried out only by professional and competent personnel in this field.

Figure 12.1
Controlled Demolition of a Multistoried Structure. *Courtesy/with permission from: The Loizeaux family & Controlled Demolition Inc., Phoenix, Maryland, USA.*

This type of demolition is known to be carried out in Europe and the United States for several years. In India, it is now becoming very prevalent. Figure 12.1 depicts the controlled demolition of Biltmore Hotel in Oklahoma City, USA, in the year 1977. Note that the collapse is inward, i.e., directed toward the center of the structure. As seen in the last photograph, hardly any major debris is noticed outside the perimeter wall of the building after its collapse.

12.3 Air Bags

Air bags have become a primary safety device in automobiles today. They complement with the seat belt and save the life of the driver in case of a crash. How the air bag saves his life is given in the following picture:

In the case of Figure 12.2(a), where the automobile is not equipped with an air bag, when there is a crash, the body (mostly the chest/ribs area) hits the steering wheel directly. The force of impact is of high order (depending on the momentum of the vehicle when it crashes), whereas the area of the human body (chest/ribs) receiving the impact is quite

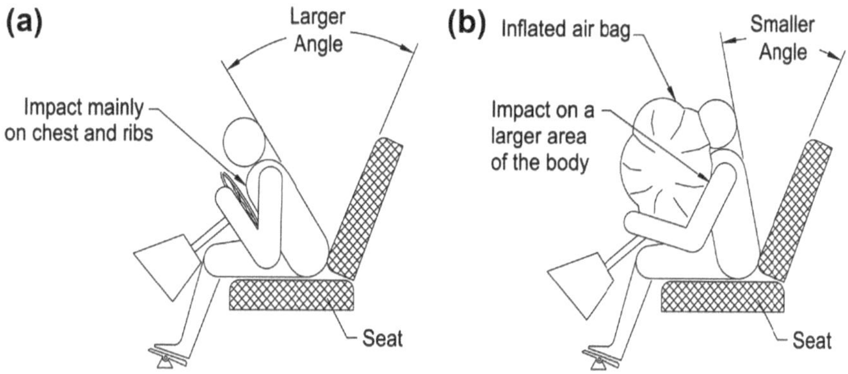

Figure 12.2
(a) Automobile Without Air Bag. (b) Automobile Equipped with Air Bag.

less. Such a high ratio of impact/area immediately kills the driver. In case of Figure 12.2(b), the automobile has been fitted with an inflatable air bag just at the center of the steering wheel. The uninflated air bag contain gas-generator HEMs, mostly a mixture of azides (like NaN_3), an oxidizer (e.g., KNO_3), and other ingredients (like SiO_2). When there is a crash, a crash sensor sends an electrical signal that ignites an initiator (Figure 12.3). The initiator ignites the gas generator mixture at the time of impact producing large volumes of nitrogen in *less than 0.05 s*, and this inflates the air bag faster than the movement of the driver's body toward the steering wheel. When the body is restrained by an inflated air bag, the force of impact is distributed over a much larger area of the body (including face and hands) resulting in less severe injuries. There is a mechanism by which the air bag gets deflated within a second after saving the life of the driver. It has been estimated that the fatality in automobile accidents has been reduced by more than 60% due to the seat belt/air bag combination.

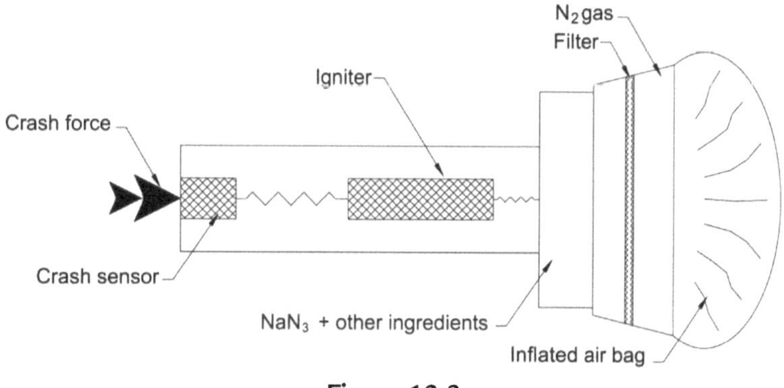

Figure 12.3
Components of Air Bag System (Schematic).

The reactions involved in an NaN_3 based gas generator are given below.

$$2NaN_3 \rightarrow 2Na + 3N_2$$
$$10Na + 2KNO_3 \rightarrow K_2O + 5Na_2O + N_2$$
$$K_2O + Na_2O + SiO_2 \rightarrow K_2Na_2SiO_4$$

Due to the toxicity and possible risks involving NaN_3, research is on to develop alternative HEMs/gas generators. A few of the potential candidates are:

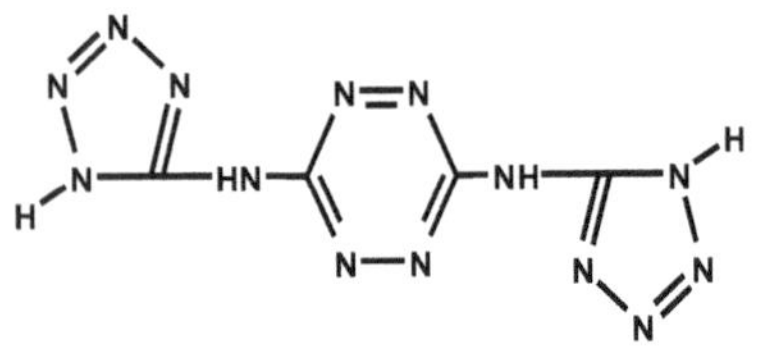

BTATz

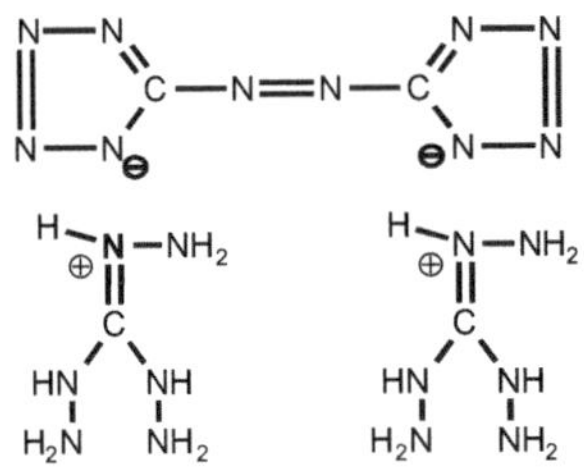

TAGAT

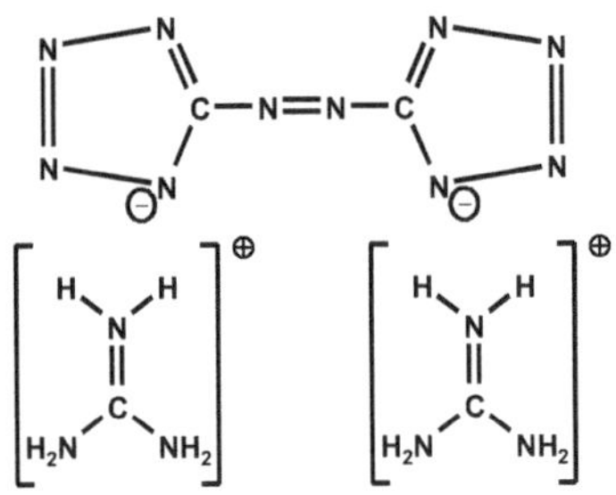

GAT

Where

BTATz	Bis tetrazolylaminotetrazine (BTATz)
TAGAT	Triamino guanidinium azotetrazolate
GAT	Guanidinium azotetrazolate

12.4 Explosive Welding

Welding of certain dissimilar metals or alloys by conventional welding is a challenging task and often impossible. This, however, has been made possible by the process of explosive welding (see Figure 12.4(a)–(c)). Let us say that we want to weld a Ni-alloy plate (cladding plate) on to a carbon steel plate (parent plate). The surfaces of the plates are cleaned and dried and the cladding plate is kept at a predetermined inclination, as shown in the figure. A layer of plastic explosive with a detonator is embedded on the cladding plate (Figure 12.4(a)). On initiation of the explosive, the detonation pressure impacts the cladding plate (also called "flyer plate") on to the parent plate with huge impact pressures (in the range of a few millions of psi, with plate speed that may vary from 100 to 300 m/s) (Figure 12.4 (b)). The interfacial pressure of the collision exceeds the yield strength of the materials, resulting in momentary plastic deformation. This results in atom-to-atom type of bonding between both the materials, giving a perfectly welded material (Figure 12.4(c)).

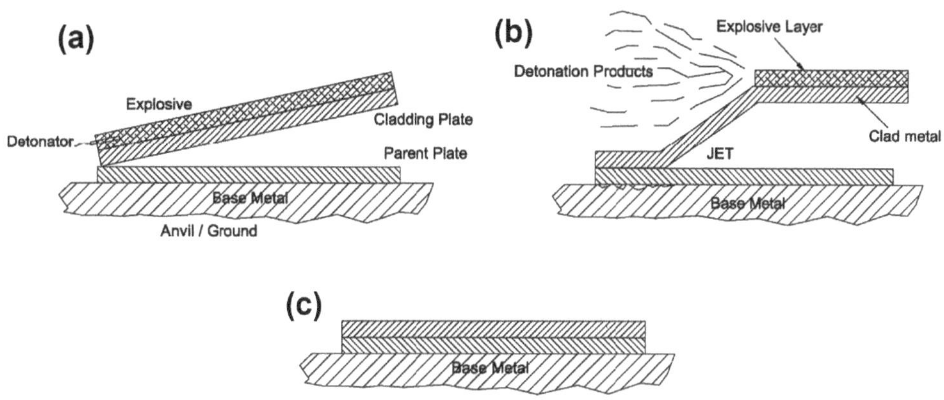

Figure 12.4
Explosive Welding of Dissimilar Metals. (a) Initial Set-up. (A Lug Support to Keep the Inclination is not Shown in the Figure). (b) After Initiation of Detonation, Huge Detonation Pressure Impacts the Cladding Plate on to the Parent Plate Resulting in Instantaneous Bonding Between the Plates. (c) Explosive-Welded Plates.

The major advantages of the method of explosive welding are:

1. Dissimilar and normally unweldable metals can be welded.
2. Can be done at room temperature in air/under water/vacuum.
3. The process is compact, portable and inexpensive.

However, there are a few disadvantages:

1. Metals/alloys should have high impact resistance and ductility.
2. The plates should have simple geometries flat/cylindrical/conical (for symmetrical travel of the shock wave).

12.5 Avalanche Control

An avalanche (a huge mass of snow and ice falling rapidly down a mountainside) often causes disasters to life and property. When snow strength (bonding between snow crystals) can no longer support its own weight, the entire mass starts sliding down causing an avalanche.

An avalanche control is a measure to intentionally trigger an avalanche using explosives (before it occurs naturally) after taking necessary precautionary measures like clearing the area from people, traffic, ski-resorts, etc. An avalanche control expert has to be conversant with mountain (snow) safety as well as explosive safety. He/she can predict the time and place of avalanche occurrence. He should be able to determine the type and quantity of explosive to be used to clear an avalanche and also the proper means of initiation.

Avalanche control prevents disasters such as people, tenements, and vehicles being buried under snow (*please refer Chapter 4, Figure. 4.14*).

12.6 Life Saving Applications

HEMs find a life-saving application for emergency exit of fighter pilots. In case the pilot wants to abandon the aircraft during an emergency, an explosive device severs and dislodges the canopy, following which a propulsion device under his seat ejects the pilot and parachute from the aircraft. The design and development of the explosive system for canopy severance and the propellants/propulsion system for seat ejection is a very critical job, as it involves the life of the pilot. Many lives have been and are being saved by a combination of seat ejection and canopy severance devices where the HEMs play a very critical role.

In the field of medicine, nitroglycerine—a well-known explosive—has saved the lives of many patients suffering from coronary heart disease. NG-based tablets are known to prevent or stop the chest pain (angina) among such patients. NG dilates the blood vessels, leading to more blood flow and oxygen supply to the heart. NG-based tablets are strictly prescribed medicines and should be taken only as per strict medical advice.

There has been an interesting and rather weird application of explosives in tenderization of meat! It was discovered by Morse Solomon, a meat scientist, and John Long, that huge quantities of meat kept under water get tenderized by subjecting them to underwater explosion. It has been estimated that this method of tenderization of meat is far cheaper than methods involving electrical power. Probably some of the potential applications by HEMs are yet to be discovered.

Suggested Reading

[1] E.G. Mahadevan, Ammonium Nitrate Explosives for Civil Applications Slurries, Emulsions and Ammonium Nitrate Fuel Oils, first ed., Wiley-VCH, 2013.

[2] The Explosive Engineer: Forerunner of Progress in Mining, Quarrying, Construction, vol. 20, Contributor Hercules Powder Company, Publisher Hercules Powder Company, 1942.

[3] E.G. Baranov, A.T. Vedin, I.F. Bondarenko, Mining and Industrial Applications of Low Density Explosives, Taylor & Francis, 1996.

[4] D.E. Davenport, Explosive Welding, American Society of Tool and Manufacturing Engineers, 1961.

[5] T.Z. Blazynski, Explosive Welding, Forming and Compaction, first ed., Springer, 1983.

[6] E.O. Paton, Explosive Welding of Metal Layered Composite Materials Welding and Allied Processes, International welding Association, 2003.

[7] R.A. Patterson, Fundamentals of Explosion Welding, ASM Handbook, vol. 6, Welding, Brazing, and Soldering (ASM International), 1993.

[8] B. Crossland, Explosive Welding of Metals and its Application, Clarendon Press, 1982.

Questions

1. What are the important factors to be considered in the controlled demolition of high-rise structures?
2. How does an automobile air bag work?
3. What is an avalanche and how can it be controlled using explosives?
4. What do you understand by canopy severance system? How does an explosive and propellant system save the life of an aircraft pilot during an emergency?
5. What is meant by explosive welding? What are its advantages over conventional welding?
6. How does nitroglycerine help in relieving angina?

Index

Note: Page numbers followed by "f" and "t" indicate figures and tables respectively

A

Activation energy, 8, 20–21, 72–74, 73f
ADN. *See* Ammonium dinitramide
Airbag, 214–216
Airblast overpressure, 167, 167t
Ammonium dinitramide (ADN), 153–154, 187–188, 189f, 203–204
Ammonium nitrate, 94–95
Ammonium perchlorate, 9–10, 154, 169, 187, 188f
Angina and NG, 218
Antacids, 128
Arrhenius equation, 73
Auxoploses, 74–75
Avalanche control, 99–100, 217–218

B

Ball powder, 10, 115, 121
Ballistics, 37, 116, 142–145
Ballistite, 4–5, 15t
Becker-Kistiakowsky-Wilson method (BKW method), 64–65
Bipropellant, 136, 137f
Bis-(5-nitro-2H-tetrazolato-N^2) tetramine cobalt(III) perchlorate (BNCP), 195, 196f
BKW method. *See* Becker-Kistiakowsky-Wilson method
Blast wave, 91–94
Blasting agents, 97–100
Blasting gelatin, 4–5, 125
BNCP. *See* Bis-(5-nitro-2H-tetrazolato-N^2) tetramine cobalt(III) perchlorate
Bomb calorimeter, 28–29
Brisance, 37, 45, 87–88
Burn rate catalysts, 12t, 143–144, 149, 153
Burning rate coefficient, 116, 142

C

C-J pressure, 58f
Calorimetric value, 28, 143, 147–148, 148t, 181
Canopy severance, 218
Cartridge case, 10, 45–46, 105–106, 108–110, 112–113, 117, 130
CD nozzle. *See* Convergent-Divergent Nozzle
Chamber Pressure, 134, 141–143
Characteristic velocity, 144–145
Charge diameter, 78, 90–91
China Lake-20 (CL-20), 59t, 184, 184t, 186f, 188, 197
Chromatography, 181–183
CL-20. *See* China Lake-20
Closed vessel test, 118
CMDB propellant. *See* Composite modified double-base propellant
Compatibility assessment, 189
Composite modified double-base propellant (CMDB propellant), 148
Composite propellants, 9–10, 12t, 142–143
Compression wave, 77, 89, 89f
Conjugated double bonds, 184
Controlled demolition, 212–214
Convergent-Divergent Nozzle (CD nozzle), 133–134
Coronary heart disease and NG, 218

D

DDT. *See* Deflagration-to-Detonation Transition
Decoppering agents, 130
Decoy flares, 158
Deflagration, 52–56, 55t
Deflagration-to-Detonation Transition (DDT), 55, 67
Delay composition, 13, 157–158
Demilitarization, 206–207
Detection of Explosives, 173–179
Detonation, 52–54
Detonation Pressure, 61–65
Detonation temperature, 37–39, 78
Detonation wave, 55–61, 67, 78–79
Diamagnetism based detector, 177
1,1-diamino-2,2-dinitroethylene (FOX-7), 199

Differential Scanning Calorimetry (DSC), 187–189, 190f
Differential Thermal Analysis (DTA), 170, 187–188
2,4-dinitroanisole (DNAN), 182–183, 191f, 202, 203t
Double base propellant, 9–10, 12t, 121, 142–143, 207
DSC. *See* Differential Scanning Calorimetry
DTA. *See* Differential Thermal Analysis

E

ECD. *See* Electron capture detector
Eco-friendly oxidizers, 153, 202–204
Eco-friendly primary explosives, 195–196
Electron capture detector (ECD), 174–175
Emulsion explosives, 94–95, 99
Energetic binders, 205–206
Energetic plasticizers, 127, 207, 208t
Energy of formation, 44
Entropy, 71
EOS. *See* Equations of state
Equations of state (EOS), 64, 117–118
Erosive burning, 144
Exhaust Gas Pressure, 135
Exhaust velocity, 139
Expansion ratio, 107, 111–112
Explosive Storage Houses, 3–4
Explosive train, 81–87, 86f
Explosive welding, 216–217
Explosives, 6–7, 71–104
Explosophores, 8, 74

F

False alarms, 176, 179
Field ion spectrometer, 177
Flame temperature, 37–39
Flash suppressants, 121, 129
Force constant, 43, 111–112, 116, 122t, 141–142
Fourier transform IR (FTIR), 185
FOX-7. *See* 1,1-diamino-2, 2-dinitroethylene
Fragmentation, 87–88, 91
Free energy, 87–88, 91
Friction Sensitivity, 170, 192, 195
FTIR. *See* Fourier transform IR

G

Gas expansion effect, 77
Gas generator composition, 214–216
Gas volume, 42
Gelatine explosives, 97
Glyceryl trinitrate (NG), 3
Gun propellant, 10–11, 105–132
Gunpowder, 1–6

H

Hazard evaluation, 170, 186–187
Heat content or enthalpy, 22
Heat of combustion, 27–29, 32, 48
Heat of explosion, 27–29
Heat of formation, 23–27, 33–34, 44, 207–209
Heat of reaction, 23
Heat Resistant Explosives, 196–197
HESH ammunition, 89
Hess's law, 24, 24f
High density, high VOD explosives, 197–199
High energy materials, 16t–17t, 19–20
High Performance Liquid Chromatography (HPLC), 183–184
HMX, 9f, 34f, 84–86, 89, 102t, 183–184, 209
HNF. *See* Hydrazinium nitroformate
HPLC. *See* High Performance Liquid Chromatography
Hugoniot curve, 58f, 59
Hydrazinium nitroformate (HNF), 152–153, 203–204
Hydrogen bonding, 84, 127

I

IEDs. *See* Improvised Explosive Devices
Igniter composition, 157–158
Illuminating composition, 158, 160
Impact Sensitivity, 170, 181, 192
Impetus, 43, 111–112
Improvised Explosive Devices (IEDs), 173, 174t
Impulse, 43, 92–93, 138–139
IMs. *See* Insensitive Munitions
IMS. *See* Ion mobility spectrometer
Incendiary composition, 158
Industrial explosives, 94–100
Insensitive Munitions (IMs), 199
Ion mobility spectrometer (IMS), 175–176
IR absorption, 184–185, 185t
Isochoric flame temperature, 37, 80, 112

K

Kieselghur, 4–5

L

Lead azide, 9f, 22, 82, 83t, 195
Lead free initiators, 195
Linear burning rate (LBR), 54–55, 110, 116–117, 142–144
Liquid oxygen, 136
Loading density, 28–29, 64, 78, 118–119
Low explosives, 6, 8, 19
Low vulnerability ammunition (LOVA), 121
Low vulnerability explosive (LOVEX), 174t

M

Marsh gas, 7, 95
Mass burning rate, 54–55, 110, 142
Mass fire, 166t, 167
Mean molar heat capacity, 39
MEMS. *See* Micro electro mechanical system
Mercury fulminate, 9, 82, 83t, 97, 195

Micro electro mechanical system (MEMS), 178
Microballoons, 98
Mining, 7, 78, 211
Molar internal energy, 39, 39t
Monopropellant, 135–136
MTNI. *See* *N*-methyl-2,4, 5-trinitroimidazole

N

N-methyl-2,4,5-trinitroimidazole (MTNI), 190–191, 202
NC. *See* Nitrocellulose
Neutral burning, 114, 115f
NG. *See* Glyceryl trinitrate; Nitroglycerin
NG tablet, 218
Nickel hydrazine nitrate (NHN), 195–196
Nitrocellulose (NC), 2–3, 4f, 31, 53, 123–124, 181
Nitroglycerin (NG), 2–3, 3f, 29f, 53, 121, 135, 207
Nitroguanidine (picrite), 9–10, 76, 85t, 121, 129
3-nitro-1,2,4-triazole-5-one (NTO), 199, 200t
NMR. *See* Nuclear magnetic resonance
NQR detector. *See* Nuclear quadrupole resonance detector
NTO. *See* 3-nitro-1,2,4-triazole-5-one
Nuclear magnetic resonance (NMR), 177, 185–186
Nuclear quadrupole resonance detector (NQR detector), 177–178

O

Obscuration, 13, 158
Octanitrocubane (ONC), 5, 5f, 195
Oil well perforation, 212
Outside Quantity Distance (OQD), 170
Overexpanded nozzle, 135
Oxygen balance (OB), 29–39, 31f, 34f, 35t

P

PBX. *See* Plastic bonded explosives
Pentaerythritol tetranitrate (PETN), 25–26, 26f, 39–40, 175
Permitted explosives, 95, 97
PETN. *See* Pentaerythritol tetranitrate
Picric acid, 75, 80–81, 85t
picrite. *See* Nitroguanidine
PIQD. *See* Process Inside Quantity Distance
Plastic bonded explosives (PBX), 101–102, 102t
Platonizers, 149
Polynitrogen caged compounds, 207–209
Prills, 34–35, 98
Primary explosives, 3, 6, 22, 24, 81–82, 83t, 195–196
Process Inside Quantity Distance (PIQD), 170
Progressive burning, 113–116
Propellant charge mass, 107, 111
Propellants, 5–6, 8–12, 10f, 19, 25, 28, 105, 110, 121, 128, 142–143, 202–207
Protective garments, 169
Pyrotechnics, 11–15, 157, 159–163, 169

Q

QD Concept, 170–171
Quarrying, 2, 99–100, 211–212

R

RDX. *See* Research and development explosive
Red Fuming Nitric acid (RFNA), 136
Reduced sensitivity research and development explosive (RSRDX), 199
Regressive burning, 115, 114f, 130
Relative force (RF), 119
Relative Front (R_f), 182
Relative vivacity (RV), 119
Research and development explosive (RDX), 6, 31, 44, 52, 76, 91, 101, 111, 173, 184, 197, 198t, 199
RF. *See* Relative force
R_f. *See* Relative Front
RFNA. *See* Red Fuming Nitric acid
Rocket motor, 133–134, 134f, 150
Rocket propellant, 11, 12t, 25, 114, 116, 133–136, 141–148, 148t, 153
RSRDX. *See* Reduced sensitivity research and development explosive
RV. *See* Relative vivacity

S

Safety directives, 168–172
Scabbing effect, 87, 88f, 89
Seat ejection, 218
Secondary explosives, 6, 83–86
Semigelatine explosives, 97
Shaped charge, 6–7, 78, 87, 89–91
Shock wave, 41, 54–59
Signal composition, 158
Single base propellant, 9–10, 121–122, 127–128
SIQD. *See* Storage Inside Quantity Distance
Slurry Explosives, 94–95, 98
Smoke composition, 159–160
Smokeless powder, 4–5, 10–11, 105
Spark sensitivity, 170, 192–193
Specific energy, 43, 79–80
Specific impulse, 43, 138–139, 147, 203
Spectroscopy, 76, 181, 184–186
Storage Inside Quantity Distance (SIQD), 170–171
Surface moderants, 130

T

TACOT, 200t–201t
Taggants, 174–175

TATB. *See* Triamino trinitrobenzene
Tenderization of meat, 218
Tension wave, 89
Tetryl, 84–86, 191–192
TGA. *See* Thermogravimetric analysis
Thermal analysis, 181, 187
Thermally stable explosive, 196–197, 200t–201t, 202
Thermite composition, 162
Thermogravimetric analysis (TGA), 187, 189–191
Thermoredox detector, 176
Throat area, 146
Thrust coefficient, 144
TNAZ. *See* 1,3,3-trinitroazetidine
TNT. *See* Trinitrotoluene
Total impulse, 138
Total thrust, 134, 145
Toxic Hazards, 169–170
Tracer composition, 158
Triamino trinitrobenzene (TATB), 84, 200t–201t
1,3,3-trinitroazetidine (TNAZ), 202
Trinitrotoluene (TNT), 3, 9f, 30–31, 54, 63, 100, 101t, 199–202
Triple base propellant, 9–10, 121, 129

U

Hazard Classification, 166–167
Underexpanded nozzle, 135
Unit of I_{sp}, 138

V

Velocity of detonation (VOD), 6–7, 36, 54, 59, 62, 77–79, 85t, 181
Vielle Law, 53, 116–118, 142
Vivacity, 116, 119

W

Waste Disposal, 163, 167, 172

Kolibri 360 GmbH
Ferdinand-Jühlke-Straße 7
99095 Erfurt, Deutschland
produktsicherheit@kolibri360.de